Application of Soft Computing Techniques in Mechanical Engineering

This text covers the latest intelligent technologies and algorithms related to the state-of-the-art methodologies of monitoring and mitigation of mechanical engineering. It covers important topics, including computational fluid dynamics for advanced thermal systems, optimizing performance parameters by Fuzzy logic, design of experiments, numerical simulation, and optimizing flow network by artificial intelligence.

The book:

- Introduces novel soft computing techniques needed to address sustainable solutions for the issues related to materials and manufacturing process.
- Provides perspectives for the design, development, and commissioning of intelligent applications.
- Discusses the latest intelligent technologies and algorithms related to the state-of-the-art methodologies of monitoring and mitigation of sustainable engineering.
- Explores future-generation sustainable and intelligent monitoring techniques beneficial for mechanical engineering.
- Covers implementation of soft computing in the various areas of engineering applications.

This book introduces soft computing techniques in addressing sustainable solutions for the issues related to materials and manufacturing process. It will serve as an ideal reference text for graduate students and academic researchers in diverse engineering fields, including industrial, manufacturing, thermal, fluid, and materials science.

Computational and Intelligent Systems Series

In today's world, the systems that integrate intelligence into machine-based applications are known as intelligent systems. In order to simplify the man–machine interaction, intelligent systems play an important role. The books under the proposed series will explain the fundamentals of intelligent systems, review the computational techniques, and offer step-by-step solutions of the practical problems. Aimed at senior undergraduate students, graduate students, academic researchers, and professionals, the proposed series will focus on broad topics, including artificial intelligence, deep learning, iImage processing, cyber physical systems, wireless security, mechatronics, cognitive computing, and industry 4.0.

Application of Soft Computing Techniques in Mechanical Engineering
Amar Patnaik, Vikas Kukshal, Pankaj Agarwal, Ankush Sharma
and Mahavir Choudhary

Application of Soft Computing Techniques in Mechanical Engineering

Edited by
Amar Patnaik
Vikas Kukshal
Pankaj Agarwal
Ankush Sharma
Mahavir Choudhary

CRC Press
Taylor & Francis Group
Boca Raton London New York

CRC Press is an imprint of the
Taylor & Francis Group, an **informa** business

First edition published 2023
by CRC Press
6000 Broken Sound Parkway NW, Suite 300, Boca Raton, FL 33487-2742

and by CRC Press
4 Park Square, Milton Park, Abingdon, Oxon, OX14 4RN

CRC Press is an imprint of Taylor & Francis Group, LLC

Library of Congress Cataloging-in-Publication Data
Names: Patnaik, Amar (Mechanical engineer), editor.
Title: Application of soft computing techniques in mechanical engineering /
edited by Amar Patnaik, Vikas Kukshal, Pankaj Agarwal, Ankush Sharma,
and Mahavir Choudhary.
Description: First edition. I Boca Raton : CRC Press, 2023. I Series:
Computational and intelligent systems series I Includes bibliographical
references and index.
Identifiers: LCCN 2022030762 (print) I LCCN 2022030763 (ebook) I ISBN
9781032191027 (hbk) I ISBN 9781032191034 (pbk) I ISBN 9781003257691 (ebk)
Subjects: LCSH: Mechanical engineering--Data processing. I Soft computing.
Classification: LCC TJ153 .A69 2023 (print) I LCC TJ153 (ebook) I DDC
621.0285--dc23/eng/20221012
LC record available at https://lccn.loc.gov/2022030762
LC ebook record available at https://lccn.loc.gov/2022030763

ISBN: 978-1-032-19102-7 (hbk)
ISBN: 978-1-032-19103-4 (pbk)
ISBN: 978-1-003-25769-1 (ebk)

DOI: 10.1201/9781003257691

Typeset in Sabon
by SPi Technologies India Pvt Ltd (Straive)

Contents

Preface

Soft computing has played an important role not only in theoretical paradigms but also in the design of a wide range of mechanical systems. The book entitled "Application of Soft Computing Techniques in Mechanical Engineering" covers a comprehensive range of soft computing techniques applied in various fields of Mechanical Engineering. This book offers state-of-the-art results from ongoing research works being undertaken in various research laboratories and educational institutions, with a focus on soft computing techniques.

All the chapters were subjected to the peer-review process by the researchers working in the relevant fields before final selection. The chapters were selected based on their quality and their relevance to the book. The book will enable researchers working in the field of soft computing applied to Mechanical Engineering to explore the various areas of research and educate future generations.

The objective of the book is to investigate the growing and current cutting-edge technology in the research domains of soft computing, as well as its original, novel, and revolutionary real-world applications in the modern era. The prospective audience of the book will be academicians, engineers, decision-makers, researchers, scientists, and industrialists working in the diverse areas of Mechanical Engineering.

Editors:

Amar Patnaik
Vikas Kukshal
Pankaj Agarwal
Ankush Sharma
Mahavir Choudhary

Acknowledgements

Successful completion of this book is the outcome of the efforts of many people involved in the whole process. It is very important to acknowledge their contribution in shaping the structure of the book. It includes all the authors, reviewers, publisher, and the concerned Institutes of the editors. To begin with, all of the editors would like to thank everyone who has helped and contributed to this book with their efforts, research, and support. We thank all of the authors of the accepted chapters for allowing their work to be transformed into book chapters. We thank all of the authors for their unwavering support throughout the lengthy review process.

We would like to acknowledge all the reviewers associated in the review process of all the chapters. Their continuous support and prompt completion of the review process have helped the editors to complete the book within the stipulated time. Their valuable suggestions and direction aided in the improvement of the quality of the chapters chosen for publishing in the book.

It is very difficult to express gratitude to everyone associated with the book. Therefore, we would like to acknowledge all the persons involved in the completion of the book directly or indirectly. Last but not the least, we would like to thank our family members for extending their support in allowing us to complete the book.

Editors:

Amar Patnaik
Vikas Kukshal
Pankaj Agarwal
Ankush Sharma
Mahavir Choudhary

Editors

Dr. Amar Patnaik is Associate Professor of Mechanical Engineering at Malaviya National Institute of Technology Jaipur, India. He has more than 15 years of teaching experience and has taught a broad spectrum of courses at both the undergraduate and graduate levels. He has also served in various administrative functions, including Dean International Affairs, and has been a coordinator of various projects. He has guided 25 PhDs and several M. Tech theses. He has published more than 250 research articles in reputed journals, contributed five book chapters, edited one book. He has filedseven patents and granted one patent in his credit. He is also the guest editor of various reputed international and national journals. Dr. Patnaik has delivered more than 30 Guest lecturers in different institutions and organizations. He is a life member of Tribology Society of India, Electron Microscope Society of India, and ISTE.

Dr. Vikas Kukshal is presently working as an Assistant Professor in the Department of Mechanical Engineering, NIT Uttarakhand, India. He has more than 10 years of teaching experience and has taught a broad spectrum of courses at both the undergraduate and postgraduate levels. He has authored and co-authored more than 26 articles in journals and conferences and has contributed seven book chapters. Presently, he is a reviewer of various national and international journals. He is a life member of Tribology Society of India, The Indian Institute of Metals, and The Institution of Engineers. His research area includes material characterization, composite materials, high entropy materials, simulation, and modeling.

Mr. Pankaj Agarwal is presently working as Assistant Professor at the Department of Mechanical Engineering at Amity University Rajasthan, Jaipur, India. He received his M. Tech degree in Mechanical Engineering specialization from Jagannath University, Jaipur, India, in 2013 and B.E. in Mechanical Engineering from University of Rajasthan, Jaipur, India in 2007. He has published more than 20 research articles in national and international journals as well as conferences and two book chapters for international publishers. He has handled/handling journals of international repute such as Taylor & Francis Taru Publication as a guest editor and edited one book. He has organized several international conferences, FDPs, and workshops as a core team member of the organizing committee. His research interests are optimization, composite materials, simulation, and modeling and soft computing.

Dr. Ankush Sharma is presently working as a scientific officer in the Centre of Excellence for Composite Materials at ATIRA, Ahmedabad. He has completed Ph.D. in composite material from Malaviya National institute of Technology Jaipur. He has more than 6 years of teaching as well as research experience and taught a broad spectrum of courses at both the undergraduate and graduate levels. He has published more than 15 research articles in national and international journals as well as conferences and three book chapters for international publishers. He is handling journal of international repute such as Bentham Science as a guest editor and edited one book in his credit. **Dr. Sharma** has also filed one patent. He has received best research award by Institute of Technical and Scientific Research (ITSR Foundation Award-2020), Jaipur, in the year 2020. His research interests are optimization, composite materials, and tribology. He is life member of The Institution of Engineers (India).

Dr. Mahavir Choudhary is Director of Vincenzo Solutions Private Limited incubated at MNIT Innovation & Incubation Centre, Jaipur, India He received his Masters of engineering from SGSITS Indore in Production Engineering with specialization in computer integrated manufacturing (CIM). He has more than 10 years of teaching experience and has taught a broad spectrum of courses at both the undergraduate and graduate levels. He has published more than ten research articles in national and international journals as well as conferences and edited one book in his credit. His research interests are optimization, composite materials, numerical simulation, and soft computing.

Contributors

Abhimanyu
G.B. Pant University of Agriculture
and Technology
Pantnagar, Uttarakhand, India

R.R. Arakerimath
G.H. Raisoni College of
Engineering and Management
Pune, India

Abhishek Arora
National Institute of Technology
Kurukshetra
Kurukshetra, Haryana, India

Amit Arora
Malaviya National Institute of
Technology Jaipur
Rajasthan, India

Anshul Kumar Bansal
Malaviya National Institute of
Technology Jaipur
Rajasthan, India

G. K. Chhaparwal
Rajasthan Institute of Engineering
& Technology
Jaipur, Rajasthan, India

Ram Dayal
Malaviya National Institute of
Technology
Jaipur, Rajasthan, India

Satish Kumar Dewangan
National Institute of Technology,
Raipur
Chhattisgarh, India

Amit Kumar Dhiman
Indian Institute of Technology
Roorkee
Roorkee, India

Akshay Nitin Dorle
National Institute of
Technology
Silchar, India

Mohak Gaur
Malaviya National Institute of
Technology Jaipur
Rajasthan, India

Puneet Singh Gautam
G. B. Pant University of Agriculture
& Technology
Pantnagar, Uttarakhand, India

Manoj Kumar Gopaliya
The NorthCap University
Gurugram, India

Pankaj Kumar Gupta
Guru Ghasidas University
Bilaspur, India

V. K. Gupta
G. B. Pant University of Agriculture
  & Technology
Pantnagar, Uttarakhand, India

Shubam Khajuria
Punjab Engineering College
  (Deemed to be University),
Chandigarh, India

C. V. Anil Kumar
Indian Institute of Space Science
  and Technology,
Thiruvananthapuram, India

Gaurav Kumar
G.B. Pant University of Agriculture
  and Technology,
Pantnagar, Uttarakhand, India,

Manish Kumar
Malaviya National Institute of
  Technology Jaipur
Rajasthan, India

Manoj Kumar
G.B. Pant University of Agriculture
  and Technology
Pantnagar, Uttarakhand, India

Rajesh Kumar
National Institute of Technology
  Kurukshetra
Kurukshetra, Haryana, India

Ravi Kumar
Indian Institute of Technology
  Roorkee, India

Pallab Sinha Mahapatra
Indian Institute of Technology
  Madras
Chennai, India

Siddhant Mohapatra
Indian Institute of Technology
  Madras
Chennai, India

D.S. Murthy
G.B. Pant University of Agriculture
  and Technology
Pantnagar, Uttarakhand, India

Aishwarya Narang
Center of Excellence in Disaster
  Mitigation & Management
Indian Institute of Technology
  Roorkee
Roorkee, India

Subash Chand Pal
Malaviya National Institute of
  Technology Jaipur
Rajasthan, India

Saurabh Pandey
Centre for Energy and Environment
  Engineering, National Institute of
  Technology
Hamirpur, India

Sahithya Pandula
Indian Institute of Technology
  Madras
Chennai, India

Gaurav Rana
Centre for Energy and Environment
  Engineering, National Institute of
  Technology,
Hamirpur, India

Sanjeev Ranjan
National Institute of
  Technology
Silchar, India

Sawan Kumar Rawat
Central University of Rajasthan
Ajmer, Rajasthan, India

Pradip Deb Roy
National Institute of Technology,
Silchar, Assam, India

Sabyasachi
Centre for Energy and Environment
Engineering, National Institute of
Technology,
Hamirpur, India

Nilesh Kumar Sharma
National Institute of Technology,
Raipur
Chhattisgarh, India

Prakash Shinde
G.H. Raisoni College of
Engineering and Management
Pune, India

Deepak Kumar Singh
National Institute of Technology
Silchar, Assam, India

Jogender Singh
Indian Institute of Space Science
and Technology
Thiruvananthapuram, India

N.S. Thakur
Centre for Energy and
Environment Engineering,
National Institute
of Technology
Hamirpur, India

Vikas
Punjab Engineering College
(Deemed to be University)
Chandigarh, India

Pradeep Kumar Vishnoi
G. B. Pant University of Agriculture
& Technology
Pantnagar, Uttarakhand, India

Ankit Yadav
Punjab Engineering College
(Deemed to be University)
Chandigarh, India

Surendra Singh Yadav
National Institute of
Technology
Silchar, Assam, India

Moh Yaseen
G.B. Pant University of Agriculture
and Technology
Pantnagar, Uttarakhand,
India

Chapter 1

Numerical investigation of a solar air heater duct with detached ribs

G. K. Chhaparwal
Rajasthan Institute of Engineering & Technology, Jaipur, India

Ram Dayal
Malaviya National Institute of Technology, Jaipur, India

CONTENTS

1.1 INTRODUCTION: BACKGROUND AND DRIVING FORCES

Today the world is facing two big challenges: ever-increasing hunger for the energy and degrading environmental conditions due to the consumption of primary source of energy (fossil fuels). Both challenges can be met with one solution, which is to use renewable source of energy. It is carbon-free unlike fossils and has sufficient capacity to fulfill the need for energy of the whole world. Previously, wind energy was a very favorable segment in renewables, but in recent trends solar energy has become the first-choice investment as well as research. This may be because it is available in abundance, can be installed anywhere, and can be scaled from a large GW capacity plant to 100 W small panel to run a street lamp. In the solar energy research domain, photovoltaic panel is the most investigated topic; however, solar thermal energy has applications such as cooling, heating, ventilation, drying, etc., which can reduce the burden over the consumption of electricity.

A solar air heater (SAH) is such a device that captures the solar radiation and uses it to heat the ambient air [1]. The heated air can be further used in various agricultural, industrial, commercial, and hospitality applications. However, SAHs have very low thermal efficiency [2]. This can be improved by multi-passing [3], multi-glazing [4], and applying various surface modification techniques such as fins, baffles, attached rib-roughness, corrugation, etc. Of these techniques, rib-roughness is the most commonly studied

DOI: 10.1201/9781003257691-1

technique in the SAH-based research domain. It can be further of mainly four types: transverse, inclined, V-shaped, and arc-shaped [5]. There are three major problems with these ribs: first, it is a cumbersome joining process to mount these ribs over the absorber plate; second, the number of ribs used is very large (generally placed at a pitch of 10–15 mm) which causes a large pressure drop; and third is the formation of hot zone at the contact of ribs and absorber plate that leads to lower surface heat transfer coefficient. All these issues can be resolved by using detached ribs instead of attached ribs. This simply means that the ribs can be mounted at a certain clearance to the absorber plate. It does not require any joining process; no hot zone formation and study reveals that the optimal pitch obtained is relatively higher than attached ribs which means less number of detached ribs, and hence less pressure drop [6, 7].

In the present study, a steady-state numerical investigation is carried out to see effects of various possible geometrical configurations like blockage ratio (d/H), clearance ratio (c/d), and relative longitudinal pitch (P/d) for an in-line arranged circular detached ribs over the absorber plate of a SAH for heat transfer enhancement. The results are plotted through contour plots and graphs to visualize the heat transfer and flow and predict the optimal arrangement.

1.2 NUMERICAL SETUP

In this steady-state numerical study, a two-dimensional geometry is drawn as shown in Figure 1.1 in the ANSYS FLUENT CFD tool considering the symmetry of the computational domain. The geometrical configurations like d/H, c/d, and P/d are varied as 0.2–0.32, 0.4–1.0, and 07–20, respectively, at Reynolds number 10000.

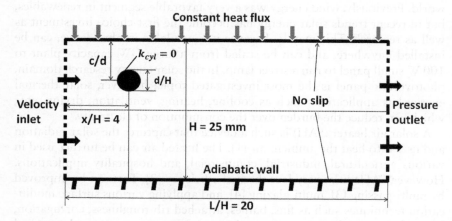

Figure 1.1 Schematic diagram of the side view of the 2D SAH-duct with circular detached rib.

After geometrical modeling, the geometry is imported in the ICEM CFD meshing tool to break this computational domain into very suitably small elements with nodes where all the governing equations will be solved using initial boundary conditions to give output parameter values. A high number of mesh elements gives more accurate results, but it also requires a very large amount of time and computing power. So mesh independence test is conducted to have an optimal mesh size that can be run at the lowest possible time and computing resources without compromising the accuracy. However, the non-uniform meshing is done in such a way that it captures the boundary layer, vortex shedding effect, and y^+ value should be less than 1 as shown in Figure 1.2.

After the completion of the meshing step, SST k-ω turbulence model is chosen as it is best for the physical phenomena that involve boundary layer effects. The boundary conditions as given in Table 1.1 are imposed over the various boundaries of the domain. The reference values are set to calculate the coefficient of drag and lift. The coupled algorithm is used for pressure-velocity coupling. The second-order scheme is used for determining pressure, momentum, turbulent kinetic energy, and turbulent dissipation rate. The convergence criteria for continuity and energy equations are kept at 10^{-6} and 10^{-9}.

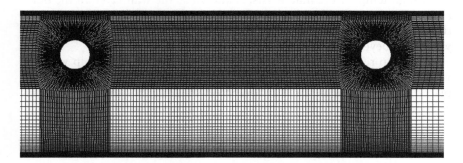

Figure 1.2 Side view of the meshing of 2D SAH-duct with detached ribs.

Table 1.1 Range of different thermal-hydraulic and geometrical parameters

Parameter	Range
Heat flux	800 W/m²
Inlet temperature	300 K
Reynolds number	10000
Blockage ratio (d/H)	0.2–0.32
Clearance ratio (c/d)	0.4–1.0
Longitudinal pitch ratio	07–20

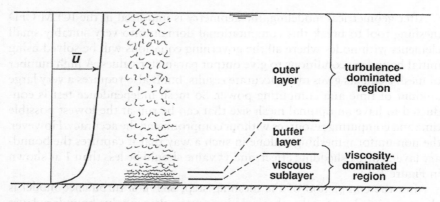

Figure 1.3 Velocity profile for laminar and turbulent flow in a confined duct and respective wall, overlap and outer layer.

The computational flow field in a confined duct can be divided into three regions: wall-layer, outer-layer, and overlap-layer, as shown in Figure 1.3. It is based on the variation of viscous force, velocity gradient, and shear stress along with the depth of the duct. Different laws govern each region. The region very near to the wall where viscous force predominates, and relatively high-temperature fluid exists, is known as the wall-layer. In the outer region, flow is inviscid and relatively cold fluid exists. The following equation gives its velocity profile. The inner and outer layer is observed below $y^+ = 5$ and beyond $y^+ = 40$, respectively.

In overlap region the U^+ values have a logarithmic relationship with the y^+ values that lie between 5 and 40. The suspended cylinders should be placed in this region as von Karman Street shedding has the ability to mix relatively hot fluid in wall-layer with cold fluid in the outer layer.

$$U^* = \frac{U}{u^*} = f(y^+) = \frac{1}{k}\ln(y^+) + B \tag{1.1}$$

$$F(\xi) = -\frac{1}{k}\ln(\xi) + B \tag{1.2}$$

where k is the von Karman constant; A and B are constants.

1.3 GOVERNING EQUATIONS

The air is working fluid with constant properties in the SAH-duct; the flow is incompressible and transitional while duct is assumed to be two-dimensional. The suspended cylinders placed in the overlap region do not have any boundary conditions and work as transverse vortex generators only.

The flow field in the SAH-duct is assumed to be two-dimensional, incompressible, transitional; and the working fluid is air, which has constant thermal-hydraulic properties. The bar does not have any boundary condition imposed (heat flux or temperature); it only generates transverse vortices. Hence, heat transfer surfaces are the same as in the smooth SAH-duct without suspended cylinders. The k is the turbulence kinetic energy and is defined as the variance of the fluctuations in velocity. The ε is the turbulence eddy dissipation, the rate at which the velocity fluctuations dissipate.

The k–ε turbulence model has few shortcomings. The stagnation and separation region over the cylinder surface has very low heat transfer due to the unrealistic production of k. Similarly, in the wake region very high intensity of k lowers the chances simulating vortex street formation. But both shortcomings are reduced by having a very fine grid. The k–ε model introduces two new variables into the system of equations. The continuity equation is then:

Continuity equation:

$$\frac{\partial(\rho u)}{\partial x} + \frac{\partial(\rho v)}{\partial y} + \frac{\partial(\rho w)}{\partial z} = 0 \tag{1.3}$$

Momentum equation:

$$\rho u \frac{\partial u}{\partial x} + \rho v \frac{\partial u}{\partial y} + \rho w \frac{\partial u}{\partial z} = -\frac{\partial p}{\partial x} + \mu \left[\frac{\partial^2 u}{\partial x^2} + \frac{\partial^2 u}{\partial y^2} + \frac{\partial^2 u}{\partial z^2} \right] \tag{1.4}$$

$$\rho u \frac{\partial v}{\partial x} + \rho v \frac{\partial v}{\partial y} + \rho w \frac{\partial v}{\partial z} = -\frac{\partial p}{\partial y} + \mu \left[\frac{\partial^2 v}{\partial x^2} + \frac{\partial^2 v}{\partial y^2} + \frac{\partial^2 v}{\partial z^2} \right] \tag{1.5}$$

$$\rho u \frac{\partial w}{\partial x} + \rho v \frac{\partial w}{\partial y} + \rho w \frac{\partial w}{\partial z} = -\frac{\partial p}{\partial z} + \mu \left[\frac{\partial^2 w}{\partial x^2} + \frac{\partial^2 w}{\partial y^2} + \frac{\partial^2 w}{\partial z^2} \right] \tag{1.6}$$

The terms given in Equations (2), (3), and (4) show acceleration, convection, pressure variation, and viscous terms in x, y, and z directions, respectively, in 3D Cartesian.

Energy equations (transport equations for k–ω model)

$$\frac{\partial}{\partial x_j}(\rho k u_j) = \frac{\partial}{\partial x_j}\left[\left(\mu + \frac{\mu_t}{\sigma_k} \right) \frac{\partial k}{\partial x_j} \right] + G_k - Y_k + S_k \tag{1.7}$$

$$\frac{\partial}{\partial x_j}(\rho \omega u_j) = \frac{\partial}{\partial x_j}\left[\left(\mu + \frac{\mu_t}{\sigma_\omega} \right) \frac{\partial \omega}{\partial x_j} \right] + G_\omega - Y_\omega + D_\omega + S_k \tag{1.8}$$

In addition to the independent variables, the density ρ, turbulent viscosity μ_t, and the velocity vector u_j are treated as known quantities from the Navier–Stokes method. G_k and G_ω is the production of turbulence kinetic energy and generation of specific dissipation rate, respectively. Y_k and Y_ω represent the dissipation of k and ω due to turbulence. S_k and S_ω are user-defined source terms. D_ω represents the cross-diffusion term.

1.4 RESULTS AND DISCUSSIONS

The variation of the velocity contour profile and temperature contour profile is shown in Figures 1.4 and 1.5, respectively. The velocity plots show that due to the presence of vortices in the wake region, the downstream flow gets disturbed. The disturbed flow makes it easy for the relatively cold fluid in the core region to mix with the relatively hot fluid nearby the absorber plate. The velocity gradient is clearly visible in the form of different colors. A similar pattern of temperature gradient is observed through the temperature contour plots. It also suggests the flow of heat from the absorber plate to the core region fluid flow. The blockage ratio d/H is increased from 0.2 to 0.32. The higher the diameter of the detached ribs, higher will be the obstruction to the overlap region flow and higher will be the heat transfer with increased friction factor as shown in Figure 1.6. When the clearance ratio c/d is increased from 0.4 to 1.0, the distance from the heated surface increases, which could lower the heat transfer enhancement while increase the strength of the vortices by increasing its size. The optimum geometrical configuration obtains around $c/d = 0.6$. The relative longitudinal pitch ratio P/d is increased from 07 to 20. Increasing the pitch ratio increases the inter space between two consecutive cylinders, which in turn increases the number of vortices. However, the strength of vortices decreases with increasing P/d; hence, an optimal tradeoff gives $P/d = 10$ as a favorable value (Figure 1.7).

1.5 CONCLUSION

The detached ribs are never studied in the SAH research domain. This study provides an opportunity to see the effect of circular shaped detached ribs over the thermal-hydraulic performance of the SAH-duct. In observation, the blockage ratio d/H, clearance ratio c/d, and relative longitudinal pitch ratio P/d are investigated in the domain of 0.2–0.32, 0.4–1.0, and 07–10, respectively. The optimal tradeoff between the various flow conditions like disturbance in the overlap region, strength and number of vortices, and distance from the heated surface give the optimal values of the geometrical configurations as $c/d = 0.6$ and $P/d = 10$, while both Nusselt number and friction increase continuously with d/H.

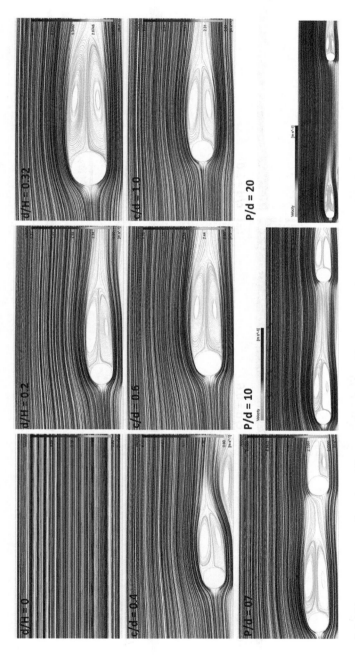

Figure 1.4 Velocity contour plots for different *d/H, c/d,* and *P/d* at *Re* = 10000.

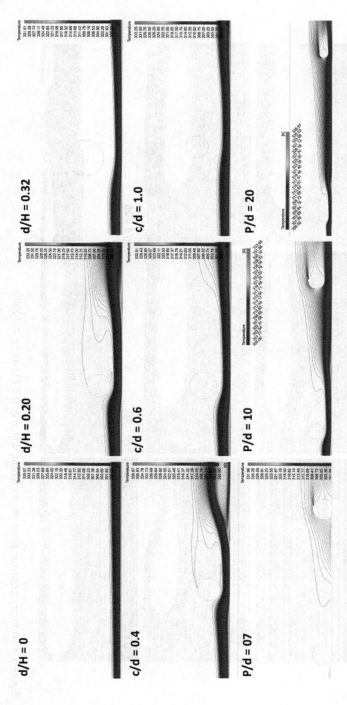

Figure 1.5 Temperature contour plots for different *d/H*, *c/d*, and *P/d* at *Re* = 10000.

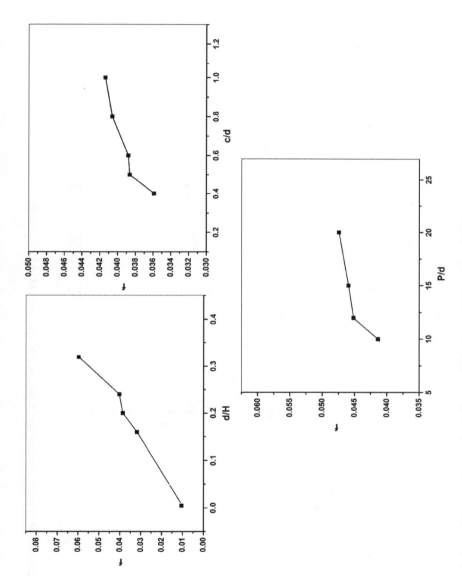

Figure 1.6 Variation of friction factor for different *d/H*, *c/d*, and *P/d* at *Re* = 10000.

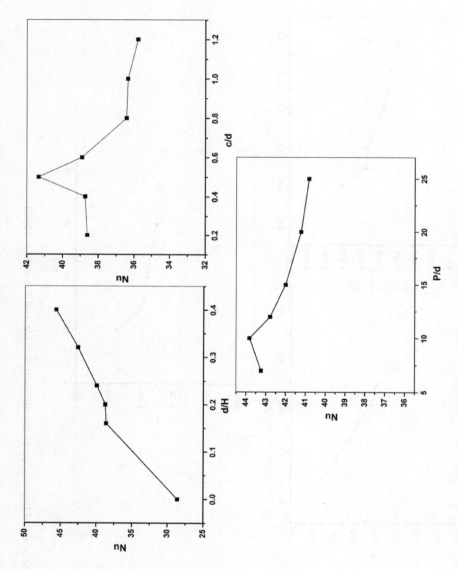

Figure 1.7 Variation of Nusselt number for different *d/H*, *c/d*, and *P/d* at *Re* = 10000.

REFERENCES

1. S. Kumar and R. P. Saini, "CFD based performance analysis of a solar air heater duct provided with artificial roughness," *Renew. Energy*, vol. 34, no. 5, pp. 1285–1291, 2009.
2. M. H. E. Ahmed and H. A. Mohammed, "An overview on heat transfer augmentation using vortex generators and nanofluids: Approaches and applications," *Renew. Sustain. Energy Rev.*, vol. 16, pp. 5951–5993, 2012.
3. H. P. Garg, V. K. Sharma, and A. K. Bhargava, "Theory of multiple-pass solar air heaters," *Energy*, vol. 10, no. 5, pp. 589–599, 1985.
4. C. Choudhury, P. M. Chauhan, H. P. Garg, W. W. Charters, P. S. Lee, S. V. Garimella, A. S. Mujumdar, T. H. Ko, C. P. Wu, M. Choi, and K. Cho, "Effect of the aspect ratio of rectangular channels on the heat transfer and hydrodynamics of paraffin slurry flow," *Int. J. Heat Mass Transf.*, vol. 21, no. 1, pp. 283–296, 1971.
5. G. Chhaparwal, A. Srivastava, and R. Dayal, "Artificial repeated-rib roughness in a solar air heater – a review," *Sol. Energy*, vol. 194, pp. 329–359, 2019.
6. G. K. Chhaparwal, A. Srivastava, and R. Dayal, "Numerical study of an asymmetrically heated rectangular duct with suspended cylinders," *MATEC Web Conf. 11th ICCHMT*, Krakow, vol. 240, p. 04002, 2018a.
7. G. Chhaparwal, A. Srivastava, and R. Dayal, "Heat transfer enhancement in a rectangular duct: Exploiting von-Karman effect," *16th Int. Heat Transf. Conf. Beijing*, pp. 5397–5416, 2018b.

Chapter 2

CFD analysis for thickness optimization of a rotating packed bed (RPB)

Gaurav Kumar, Abhimanyu and D.S. Murthy

G.B. Pant University of Agriculture and Technology, Pantnagar, India

CONTENTS

2.1 INTRODUCTION

The conventional columns providing gas–liquid interactions inside a porous packing are employed in the chemical industries to perform the processes such as absorption, adsorption, distillation, etc. These columns generally have huge size, low efficiency, high cost, and limitation in mass transfer due to gravity being the sole driving force. These flaws of conventional columns inspired the development of a rotating packed bed (RPB) which works on the high gravity principle as it uses much higher centrifugal force than the gravitational force. The centrifugal field in the RPBs can be 100–1000 times higher than the gravitational force, and it can achieve the volume reduction of 2–3 folds when compared to conventional packed bed columns [1, 2]. The RPBs have been widely established in the chemical industries owing to their tremendous industrial applications. The use of high centrifugal force has successfully lowered the flooding tendency when compared with that in a conventional column, and this increased the limits for gas and liquid flow rate at which the RPB could be operated.

The analogy between heat and mass transfer motivates to search for similar intensification in thermal systems such as cooling towers. The use of RPBs in the thermal system is not reported much in the literature. However, some recent researches have shown the results for thermal behavior and heat transfer mechanism inside the RPB, and these results could be employed to achieve the volume reduction in thermal devices [3–5].

DOI: 10.1201/9781003257691-2

13

The setup of an RPB consists of a rotor, air inlet and air outlet pipe, water distributor, electric motors, and centrifugal pump to blow the air inside the packing by providing the required pressure gradient and protective transparent casing for process visualization and to prevent the water from splashing. The rotor is the rotating component of a cylindrical shape made up by installing a porous packing between two acrylic discs. The illustrative representation of the setup of RPB is shown in Figure 2.1.

The air is blown through the air inlet pipe and with the help of a centrifugal pump. After passing through the packing, it is released to the environment through the air outlet pipe. The water is inserted inside the rotor from the inner periphery using a provision made at the eye of the rotor. After splashing out from the peripheral edge of the rotor, the splashed water gets collected at the casing floor and flows out from the water outlet pipe. The air and water come in contact with each other inside the packing from opposite directions.

The study of pressure drop inside RPB is important for its application in thermal systems, as lesser pressure drop reduces the power required to push the gas through the rotor. Studies on pressure drop have been reported by many authors since the beginning of RPB technology [6–10]. The pressure drop inside RPB was observed to follow some unexpected trends such as with the introduction of liquid the pressure drop decreased. Some authors have found the pressure drop outside packing insignificant [7]; however, in some other studies it is substantial [11]. Zheng et al. [11] performed the experiments on RPB with fixed outer packing diameter and varying inner diameter and found that the RPB with a higher inner radius has a higher pressure drop.

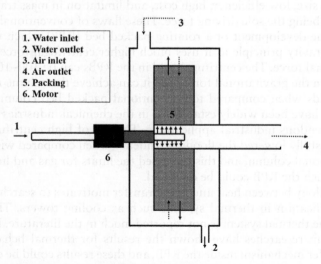

1. Water inlet
2. Water outlet
3. Air inlet
4. Air outlet
5. Packing
6. Motor

Figure 2.1 Illustrative representation of the RPB setup.

The experimental data on RPB gave essential information, but due to the rotating components and complex flow behavior inside the packing, the authors have used the computational fluid dynamics (CFD) technique [12]. In the present work, the simulation on RPB has been performed and pressure drop has been analyzed for the packing with a different inner diameter, thus optimizing the thickness of RPB by minimizing the pressure drop across the rotor.

2.2 CFD MODEL FOR RPB

The authors have used two approaches to simplify the porous medium in the past. One of them is the obstacle simplification approach in which the porous medium is described by uniformly placed simple geometries in 2D or 3D [13–15]. Although this method provides better visualization of flow inside the packing, the method requires fine grids and thus a large computational cost. Another approach is the porous media simplification model, which has been used in the present work. The geometry of the RPB was simplified by considering only three sections as shown in Figure 2.2.

Section I contained the peripheral section, section II contained the packing section, and section III contained the eye of rotor and air-outlet pipe. The air inlet was provided at the outer boundary of the peripheral section and the outlet was provided at the end of the air outlet pipe. The porous media simplification model was applied for section II by giving the values of viscous resistance coefficient, initial resistance coefficient, and porosity. The z-axis was set as the axis of rotation and the rotating sections were simulated using the multiple frame reference (MRF) model. The air was assumed

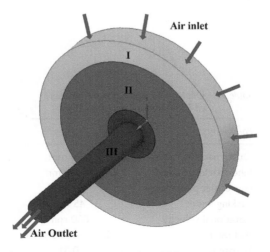

Figure 2.2 Simplified geometrical model for CFD simulations.

incompressible and the viscosity of air was considered constant during the process. The boundary condition at the inlet was set as mass flow inlet, and at the outlet it was set as pressure outlet with the outlet pressure being equal to the environmental pressure. No slip boundary condition was set at the walls.

Equations (2.1) and (2.2) are the governing equations which are general continuity and momentum balance equation, respectively, with added source term for porous media modeling.

$$\frac{\partial \rho}{\partial t} + \nabla \cdot (\rho \vec{v}) = 0 \tag{2.1}$$

$$\frac{\partial}{\partial t}(\rho \vec{v}) + \nabla \cdot (\rho \vec{v} \vec{v}) = -\nabla p + \nabla \left(\mu \left[\left(\nabla \vec{v} + \nabla \vec{v}^T \right) - \frac{2}{3} \nabla \vec{v} I \right] \right) + \rho \vec{g} + S_i \tag{2.2}$$

Here, S_i is source term for porous media modeling given in Equation (2.3).

$$S_i = -\left(\frac{\mu}{\alpha} v + \frac{1}{2} \rho C_2 v^2 \right) \tag{2.3}$$

Here, $\dfrac{1}{\alpha} = \dfrac{150(1-e)^2}{D_p^2 e^3}$ denotes viscous loss coefficient and $C_2 = \dfrac{3.5(1-e)}{D_p e^3}$ denotes inertial resistance coefficient.

The turbulence was modeled by using the standard $k - \varepsilon$ model in ANSYS Fluent. The design parameters of RPB and the operating parameters used in the simulations are provided in Table 2.1.

Table 2.1 Design and operating parameters of RPB used in the simulations

Parameters	Values
Outer packing radius	120 mm
Inner packing radius	20–60 mm
Gas outlet pipe radius	20 mm
Packing width	45 mm
Rotational velocity	800 rpm
Airflow rate	29.35–58.70 m³/h
Porosity	0.81

2.3 RESULTS AND DISCUSSION

To validate the simulation results, the experiments were performed on the rotor with an inner diameter of 80 mm. To avoid the error produced by the rotating components during experimentation, the probes to measure pressure drop were fixed on stationary parts. The pressure drop across the rotor was measured between the air inlet pipe provided at the top of the casing and the end of the air outlet pipe. The pressure drop was recorded for different gas flow rates. Figure 2.3 represents the validation curve showing the variation of pressure drop with airflow rate by using both experimental method and CFD simulations.

It can be seen from Figure 2.3 that the error in the simulations was within 15% of the experimental results. Thus, the standard $k - \varepsilon$ model in ANSYS Fluent has been proved to be suitable for predicting drop in pressure inside an RPB. The calculated values of pressure drop were less than the values obtained from experiments which may be due to neglecting the effect of water distributor in the rotor eye and pressure drop inside the casing.

The pressure drop across the RPB of the different inner radius has been shown in Figure 2.4. The pressure drop was measured as the difference between facet averaged static pressure at inlet and outlet face.

Figure 2.4 shows that for a higher inner radius, the pressure drop was higher. As the inner radius decreased, the pressure drop decreased up to a certain radius and after that, it started increasing. This phenomenon was unusual as with the increase in the packing thickness, the overall pressure drop was decreasing. Figure 2.4 also illustrates the pressure drop variation with different gas flow rates. For each RPB, irrespective of its inner radius, the pressure drop at a higher airflow rate was higher. Figure 2.5 shows the pressure contours for RPBs of different inner radius at a plane perpendicular to the rotor axis sliced from the center of the rotor. The gas flow rate was

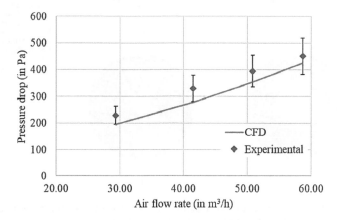

Figure 2.3 Validation of pressure drop results obtained from CFD simulation with the experimental results for different airflow rates.

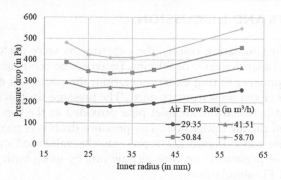

Figure 2.4 Variation of differential pressure across RPB with varying inner radius.

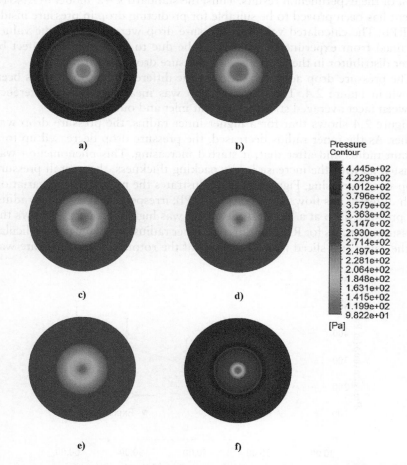

Figure 2.5 Pressure contours at a plane sliced from the center of the rotor perpendicular to rotor axis for RPB with inner radius (mm), a) 20, b) 25, c) 30, d) 35, e) 40, and, f) 60.

kept as 50.84 m³/h and the rotor speed was kept as 800 rpm. It is apparent from Figure 2.5 that pressure drop for the inner radius of 30 mm is the minimum and for 60 mm it is the maximum.

The pressure drop inside RPB can be broadly understood in two parts. One is pressure drop across the packing section and the other is pressure drop outside the packing section, i.e., the pressure drop inside the casing, rotor eye, and gas outlet pipe. The pressure drop in the different sections can be explained by identifying and studying forces acting on air in these sections. The forces acting on the air when it entered the packing were frictional force due to the interaction between air and porous media, centrifugal force due to rotation of the rotor, and Coriolis force due to radially inward movement of air in the rotating frame of the rotor.

As the air entered the eye section of the rotor it kept rotating and the effect of Coriolis force became drastic as there was no packing, therefore no drag to oppose the Coriolis force. Also, the air changed its direction toward the air outlet pipe by taking a 90° bend and a shift in momentum was observed. All these factors contributed to the pressure drop outside the packing section. The division of pressure drop in pressure drop across packing ($\Delta P_{Packing}$) and pressure drop outside packing ($\Delta P_{Outside\ Packing}$) is shown in Figure 2.6 for airflow rate 50.84 m³/h and rotor speed 800 rpm. As the thickness of the packing section was increased, i.e., the inner diameter of the packing was reduced, the pressure drop inside the packing section increased and vice versa. However, as the packing thickness was decreased, the space in the eye section increased and so did the angular velocity of air in the eye section. This caused the increase in pressure in the eye section, and thus the overall pressure drop increased.

As seen in Figure 2.6, up to a certain packing thickness, the pressure drop in the eye section was higher than that in packing section, but after the inner

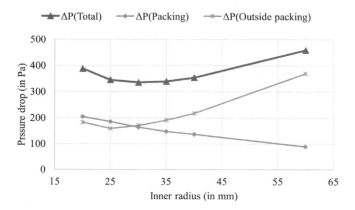

Figure 2.6 Pressure drop inside packing and outside packing for RPBs with different inner radius at airflow rate 50.84 m³/h and angular velocity 800 rpm.

radius decreased further than 30 mm, the pressure drop in the packing section became higher than the pressure drop outside the packing. At an inner radius of 30 mm, the overall pressure drop was found to be minimum.

2.4 CONCLUSIONS

The CFD simulations on the RPB were performed and the air pressure drop behavior was observed for single-phase flow. The simulations were performed in the ANSYS fluent software using standard $k - \varepsilon$ model for turbulence modeling, and the results were validated with the experiments. The error in the simulation was under 15%. The pressure drop variation with the airflow rates showed the increment in pressure drop as the gas flow rate increased. RPBs of fixed outer packing radius and varying inner radius were analyzed and it was observed that the differential pressure across the RPB increased with the increase in inner radius of packing. The pressure difference was also observed to increase when the inner radius of packing was reduced further after a certain value. Therefore, an optimized radius of 30 mm was obtained for the lowest overall pressure drop.

NOMENCLATURE

D_p	Particle diameter
e	Porosity
g	Gravitational acceleration
I	Unit stress tensor
k	Turbulence kinetic energy
p	Pressure
v	Velocity
α	Thermal diffusivity
ε	Turbulence kinetic energy dissipation rate
μ	Dynamic viscosity
ρ	Density

REFERENCES

1. D.P. Rao, A. Bhowal, and P.S. Goswami, "Process intensification in rotating packed beds (HIGEE): An appraisal." *Ind. Eng. Chem. Res.*, vol. 43, pp. 1150–1162, 2004. https://doi.org/10.1021/ie030630k.
2. L. Agarwal, V. Pavani, D.P. Rao, and N. Kaistha, "Process intensification in HiGee absorption and distillation: Design procedure and applications." *Ind.*

Eng. Chem. Res., vol. 49, pp. 10046–10058, 2010. https://doi.org/10.1021/ie101195k.

3. Saurabh, and D.S. Murthy, "Analysis and optimization of thermal characteristics in a rotating packed bed." *Appl. Therm. Eng.*, vol. 165, pp. 114533, 2020. https://doi.org/10.1016/j.applthermaleng.2019.114533.

4. Saurabh, S. Kumar, and D.S. Murthy, "Insights into thermal transactions of a novel rotating packed bed." *J. Therm. Sci. Eng. Appl.*, vol. 14, pp. 011003, 2022. https://doi.org/10.1115/1.4050836.

5. S. Kumar, G. Kumar, and D.S. Murthy, "Experimental investigation on thermal performance characteristics of rotating packed bed." *Exp. Heat Transf.*, vol. 35, pp. 1–13, 2022. https://doi.org/10.1080/08916152.2022.2027576.

6. M. Keyvani, and N.C. Gardner, "Operating characteristics of rotating beds". Technical progress report for the third quarter 1988, Pittsburgh, PA, and Morgantown, WV, 1988. https://doi.org/10.2172/10172400.

7. M.P. Kumar, and D.P. Rao, "Studies on a high-gravity gas-liquid contactor." *Ind. Eng. Chem. Res.*, vol. 29, pp. 917–920, 1990. https://doi.org/10.1021/ie00101a031.

8. P. Sandilya, D.P. Rao, A. Sharma, and G. Biswas, "Gas-phase mass transfer in a centrifugal contactor." *Ind. Eng. Chem. Res.*, vol. 40, pp. 384–392, 2001. https://doi.org/10.1021/ie0000818.

9. M. Lashkarbolooki, "A general model for pressure drop prediction across a rotating packed bed." *Sep. Sci. Technol.*, vol. 52, pp. 1843–1851, 2017. https://doi.org/10.1080/01496395.2017.1302476.

10. J.R. Hendry, J.G.M. Lee, and P.S. Attidekou, "Pressure drop and flooding in rotating packed beds." *Chem. Eng. Process. – Process Intensif.*, vol. 151, pp. 107908, 2020. https://doi.org/10.1016/j.cep.2020.107908.

11. C. Zheng, K. Guo, Y. Feng, C. Yang, and N.C. Gardner, "Pressure drop of centripetal gas flow through rotating beds." *Ind. Eng. Chem. Res.*, vol. 39, pp. 829–834, 2000. https://doi.org/10.1021/ie980703d.

12. Y.C. Yang, Y. Ouyang, N. Zhang, Q.J. Yu, and M. Arowo. "A review on computational fluid dynamic simulation for rotating packed beds." *J. Chem. Technol. Biotechnol.*, vol. 94, pp. 1017–1031, 2019. https://doi.org/10.1002/jctb.5880.

13. P. Xie, X. Lu, X. Yang, D. Ingham, L. Ma, and M. Pourkashanian, "Characteristics of liquid flow in a rotating packed bed for CO2 capture: A CFD analysis." *Chem. Eng. Sci.*, vol. 172, pp. 216–229, 2017. https://doi.org/10.1016/j.ces.2017.06.040.

14. Y. Yang, Y. Xiang, Y. Li, G. Chu, H. Zou, M. Arowo, and J. Chen, "3D CFD modelling and optimization of single-phase flow in rotating packed beds." *Can. J. Chem. Eng.*, vol. 93, pp. 1138–1148, 2015. https://doi.org/10.1002/cjce.22183.

15. T.Y. Guo, X. Shi, G.W. Chu, Y. Xiang, L.X. Wen, and J.F. Chen, "Computational fluid dynamics analysis of the micromixing efficiency in a rotating-packed-bed reactor." *Ind. Eng. Chem. Res.*, vol. 55, pp. 4856–4866, 2016. https://doi.org/10.1021/acs.iecr.6b00213.

Chapter 3

Performance evaluation of solar photovoltaic system using vertical single-axis solar tracker in Himachal Pradesh, India

Gaurav Rana, N.S. Thakur, Saurabh Pandey and Sabyasachi

Centre for Energy and Environment Engineering, National Institute of Technology, Hamirpur, India

CONTENTS

3.1 INTRODUCTION

Energy is the basic necessity for creatures to survive in the universe either in form of food or heat. The dependencies of the world on fossil fuels are a reason of worry for scientists and researchers from the past few decades. Because limited fossil fuels are about to end in the near future and fossil fuels-based energy conversion techniques are not environment friendly. So, to continue the pace of development in the world, researchers are looking for alternative fuels and renewable fuels, which will replace fossil fuels in the future. Many alternate ways to produce energy are developed, such as solar photovoltaic, solar thermal, wind energy, hydro power, fuel cell, etc., but all these alternates are yet not enough to replace fossil fuels completely in

DOI: 10.1201/9781003257691-3

23

the world because many of these are still in the development stage. Among all the renewable energy sources, solar energy is a perfect contender to fossil fuels. Using the solar photovoltaic conversion method, solar energy can be converted into electrical energy. In this process, a solar cell is used as a medium and when solar radiation falls on solar cells, it converts solar energy to electricity. The efficiency of solar photovoltaic energy conversion is quite low. The efficiency of silicon solar cells is in the range of 21%–27% [1]. Researchers have developed many types of solar cells, such as multijunction solar cells, perovskite solar cells, dye-sensitized solar cells, organic solar cells, etc., to increase efficiency [2]. Solar trackers were designed to track the path of the sun, which were later used with solar photovoltaic plants to increase the solar energy extraction of solar modules [3].

3.1.1 Solar trackers

Solar trackers are the devices that track the sun path from sunrise to sunset and can be used to rotate solar modules with respect to the sun path in order to maximize the output of the solar module. On the basis of techniques of tracking, solar trackers can be classified into active trackers, passive trackers, manual trackers and chronological trackers. Active trackers are those trackers which use any kind of sensors and microprocessors to sense the change in sun's position to control the position of a solar module [4]. Passive trackers are those which use some kind of material or fluid that can either expand or contract with changes in temperature. Hence, attaching these thermally sensitive fluids below solar modules results in different tilt angles with respect to sun's position [4, 5]. Chronological trackers are time-based trackers. In this type of tracker, the solar module rotates by a fixed amount of rotation throughout the year. Manual trackers are the trackers which are manually rotated season to season [3, 5].

On the basis of the axis of movement, solar trackers are broadly of two types: Dual-axis solar trackers and single-axis solar trackers. Dual-axis trackers are the trackers which track the sun in both horizontal and vertical axis directions. In this type of tracker, generally primary axis of rotation is the vertical axis and the secondary axis of rotation is the horizontal axis. Single-axis solar trackers track the sun path in only one direction either horizontally or vertically. Hence, the single-axis solar tracker is further divided into horizontal single-axis solar trackers and vertical single-axis solar trackers [4].

A study carried out by M. Mahendran found that using a single-axis solar tracker, 32% more electricity can be generated in Malaysia as compared to the static solar module [6]. Arian Bahrami ranked seven types of solar trackers in Africa and Europe on the basis of the Perez anisotropic model. The result shows dual-axis tracker as a top-ranked tracking strategy followed by the IEW tracking system. The ranking of the vertical axis tracker was getting improved with the increase in latitude. For latitude 60°, the vertical axis tracker extracted up to 42% of additional energy as compared to fixed modules. The author suggested a vertical axis tracker for latitude higher than 26° [7]. Guillermo Quesada performed an experiment on rooftop solar plant to investigate the tracking strategies in Canada for high latitudes and found that a dual-axis tracker can extract 25% more solar radiation as compared to a horizontal static solar module [8]. Zhimin Li presented an investigation report of the optical performance of solar panels using a vertical axis tracker and then compared it with fixed solar panels and fully tracked solar panels. An investigation was performed at 31 sites in China. It was found that optimal tilt for vertical axis tracker increases with an increase in the latitude of the site. There was 16% and 28% increment in total solar energy harvesting as compared to a fixed panel at high radiation level sites and low radiation level sites, respectively [9]. Arian Bahrami performed a study to rank the solar trackers from latitude 20°–70° and ranked all the trackers based on the Levelized Cost of Energy and energy gain. He found that with an increase in plant capacity, the vertical axis tracker went up in ranking for high latitudes [10]. Rizki Dian presented the design of a dual-axis tracker for photovoltaic panels. Four mini-PV modules were used instead of sensors and their voltage output was used to track the sun path. The result showed that 26% more energy was generated by a solar panel using dual-axis trackers [11]. Jinping Wang discussed the design and performance analysis of a single-axis tracker for parabolic trough collectors. To control the mechanical motion of the tracker, PLC is used, which can be programmed accordingly. The results have shown that a PLC-based tracking system had a tracking error of less than 0.6° [12]. Yongqiang Zhu presented a modern design of a single-axis tracker named TR axis tracker. In this design, the author placed solar panels over a wedge, which is tilted at an optimal angle for the vertical axis tracker. The tracker had a vertical axis movement setup but a wedge was placed over a sloped surface. Hence every movement changes the tilt angle in the horizontal plane also. The tilt of the slope surface can be determined by using sun-earth geometry. The results show that the TR axis tracker can harvest 96.40% of solar radiation compared to the dual-axis tracker [13]. Zakaria presented his work on dual-axis solar trackers using omnidirectional vision technology. Using a catadioptric camera, images of the sun were clicked and the sun's position information was gathered by the same. This system could track 360° horizontally and 200° vertically [15]. Shashwati Ray designed a tilted single-axis tracker and Azimuth altitude dual-axis tracker, and then compared their performances. The result has shown that the tilted single-axis tracker had a tracking error of 2.5° which is the same as the azimuth altitude dual axis tracker. But the increase in efficiency of a panel for the single-axis tracker was 30% while it was 40% for the dual-axis tracker [16]. Tralochan Kaur designed a low-cost Arduino-based two-axis tracking system and used light-dependent resistors to sense the sun's position. Using LabVIEW, the tracker system was tested and the result showed that there was a performance improvement of 13.44% as compared to the immobile system [17].

Previous work on solar tracker systems was mostly focused on dual-axis solar trackers, and the performance investigation of solar trackers in the Indian region is yet to be determined. So, in this chapter, the Hamirpur district of Himachal Pradesh, India, has been chosen for the performance evaluation of the solar tracker. On the basis of latitude of test location, previous studies suggested that the vertical single-axis solar tracker is best suited for Himachal Pradesh.

3.2 METHODOLOGY

Methodology section includes site selection, astronomical data collection and analysis, optimum tilt angle calculation and fabrication of the solar tracker. It can be better understood by the flow chart shown in Figure 3.1.

3.2.1 Astronomical data analysis of test location

Previous study has shown that a vertical single-axis sun tracker is suitable for high-latitude locations. It is preferable to gather astronomical data and prior yearly solar radiation data before developing a vertical single-axis solar tracker in order to determine the optimal tilt angle for the solar panel at a specific test location. The test site in this case was in the Hamirpur district of Himachal Pradesh, India (31.7083 N, 76.5273 E). Meteonorm 7.3 was used to acquire astronomical data, which was then analyzed using PVsyst 7.0 software.

3.2.2 Optimum tilt angles

The tilt angle of a PV panel at which the highest transposition factor may be obtained is known as the optimal tilt angle. The transposition factor is the proportion of incident irradiation on the plane to horizontal irradiation, i.e., what we gain (or lose) by tilting the collection plane. Graphs were generated using PVsyst results to calculate the ideal module tilt angle. Optimum tilt for test location was in the range of 32°–34° shown in Figure 3.2. Hence, tilt angle was taken equal to 33° for testing of solar panels. It was found that at tilt angle 33° and zero azimuth angle, transposition factor was the highest, i.e. 1.16. On increasing the tilt angle above optimum value, there is reduction in transposition factor. The lowest transposition factor of 0.77 is obtained at 90° tilt angle.

3.2.3 Design of tracker

Fabrication of rotating structure, control unit, motor unit and sensor unit are the four parts of the solar tracker design. These are explained in the subsequent sections:

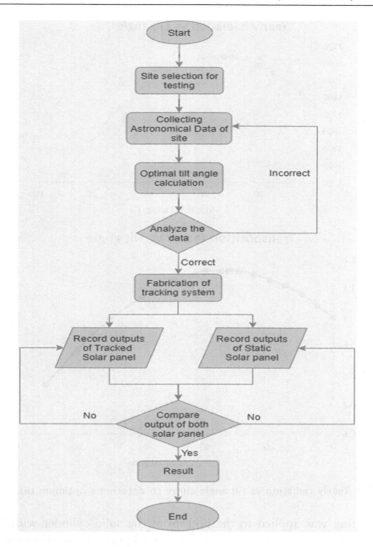

Figure 3.1 Flow chart for methodology.

1. Fabrication of rotating structure

   A hollow shaft and a solid shaft were employed to construct the building. The inner diameter of the hollow shaft was somewhat larger than the diameter of the solid shaft, and the length of the hollow shaft was longer than the length of the solid shaft. Inside the hollow shaft is a solid shaft. To avoid overloading the engine, supporting legs were welded over the hollow shaft and an H-shaped support structure for solar panels was welded over the top of the solid shaft. A thin metal

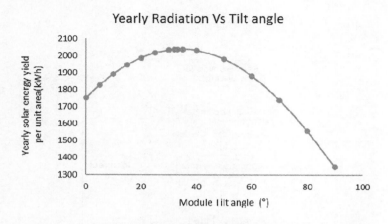

Figure 3.2 Yearly radiation vs tilt angle curve to determine optimum tilt.

coating was applied to the bottom of the solid cylinder with holes for coupling with the servo motor's metal gear. The servo motor was placed immediately below the thin metal coating on a specially built servo motor support, as shown in Figure 3.3.

2. Control unit

To control the rotation of solar tracker, Arduino uno microcontroller was used. Arduino uno was given a DC supply of 12 V using a rechargeable battery. The sensor unit provided feedback signal to Arduino and Arduino further manipulated the readings of the sensor unit and provided control input to the motor unit in order to achieve the highest possible energy production through solar panel. The flow chart for the programming of Arduino uno is shown in Figure 3.4.

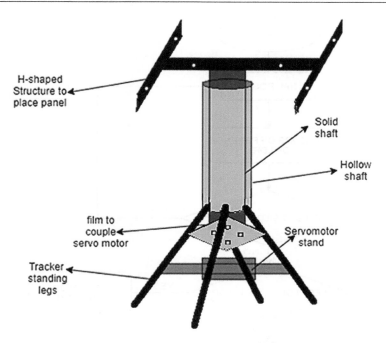

H-shaped Structure to place panel

Solid shaft

Hollow shaft

film to couple servo motor

Servomotor stand

Tracker standing legs

Figure 3.3 Rotating structure for vertical single-axis solar tracker.

3. Motor unit

For the rotation of solar tracker, MG958 servo motor was used. A servo motor is a type of motor which is given a reference signal along with power supply. According to reference signal, servo motor can control its rotation and speed. The specification of servo motor used is given in Table 3.1.

4. Sensor unit

A pair of LDRs was used as the sensor unit. LDR senses the changes in the solar radiation and produces respective signals which were fed to Arduino. LDRs were placed at the top of the solar panel. A complete image of the experimental setup fabricated for investigating the performance of the vertical axis solar tracker is shown in Figure 3.5.

3.2.4 Block diagram

Working of the vertical single-axis solar tracker can be understood through block diagram shown in Figure 3.6. Arduino board was given 12 Volt DC power while the servo motor was given 6 Volt Dc supply through rechargeable battery. The LDR sensor pair produced differential output based on the changes in solar radiation striking LDR sensors and fed it to Arduino microcontroller. Arduino uno further processed control signal and sent it to

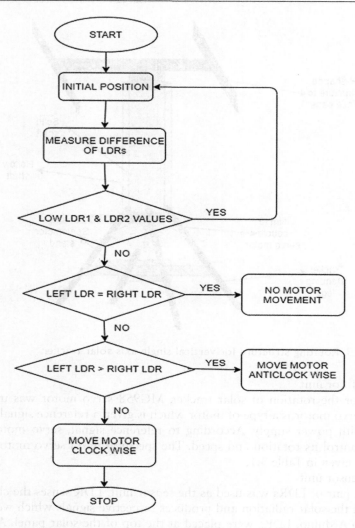

Figure 3.4 Flow chart for programming of Arduino.

Table 3.1 Specification of servo motor MG958

Operating voltage	6 Volts
Operating temperature	0–55°C
Torque	20 kg/cm
No load current	170 mA
Operating speed	0.15 sec/60°
Dimension	54.2 mm × 20 mm × 47.8 mm

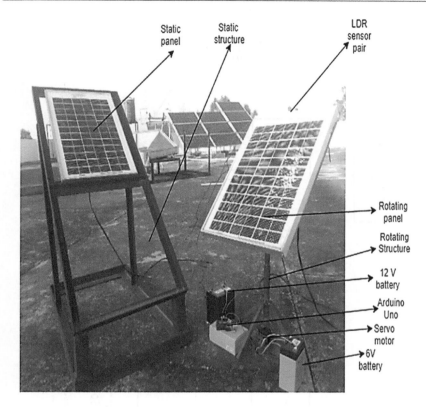

Figure 3.5 Image of the experimental setup of the vertical single-axis solar tracker.

servo motor. Servo motor rotated according to the sun's position and due to this, the tracked solar panel generated additional power with respect to the static solar panel. Output parameters of both, tracked solar panel and static solar panels, were fed to the solar analyzer and recorded.

3.2.5 Solar panel

Solar panels are the instruments which convert solar energy into electrical energy using photovoltaic conversion technique. Solar panel can be of many types such as polycrystalline silicon solar panel, monocrystalline silicon solar panel, thin film solar panel, etc. Panel used for testing was silicon poly-crystalline type. Polycrystalline solar cells are comparatively less efficient than monocrystalline solar cells because these are made of more than one type of silicon crystals. But polycrystalline solar panels are comparatively cheap in cost. Specification of solar panels used in experimental work is shown in Table 3.2.

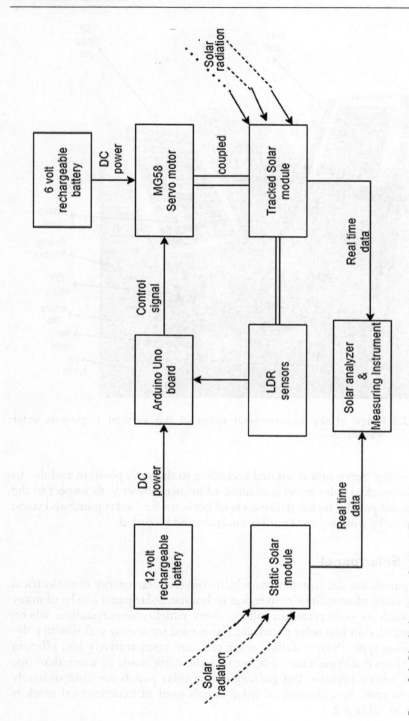

Figure 3.6 Block diagram of experiment.

Table 3.2 Specification of test solar panel at STC

Type	Silicon Polycrystalline
Area of panel	0.084 m^2
P_{max}	10 Watts
V_{oc}	21 Volts
I_{sc}	0.68 Amps
V_{mp}	16.4 Volts
I_{mp}	0.61 Amps

3.2.6 Fill factor and efficiency of test solar panel at STC

Fill factor is the ratio of product of peak current and peak voltage to the product of open circuit voltage and short circuit current. It is the measure of squareness of I-V curve of a solar cell. It is calculated by the following formula [14]:

$$\text{Fill factor, } FF = \frac{(V_m \times I_m)}{(V_{oc} \times I_{sc})}$$

$$FF = \frac{16.4 \times 0.61}{21 \times 0.68} = 0.70$$

Efficiency of a solar cell is the ratio of solar energy striking on the solar cell to the electrical energy generated by the solar cell. It is calculated by using the following formula [24]:

$$\text{Efficiency (\%)} = \frac{(V_{oc} \times I_{sc} \times FF)}{(P_{in} \times A_P)} \times 100$$

$$\text{Efficiency (\%)} = \frac{21 \times 0.68 \times 0.70}{1000 \times 0.084} = 11.9\%$$

3.3 RESULTS AND DISCUSSION

Readings of solar photovoltaic modules were taken into two categories: Cloudy sky days and clear sky days. During cloudy days readings were taken to analyze whether vertically tracked solar panel generates more or less energy as compared to the static solar panel in test location. And readings during clear sky days were recorded to determine the overall performance of the solar panel using vertical single-axis solar tracker.

3.3.1 Cloudy sky days

To determine the performance of vertical single-axis solar tracker during cloudy days, readings were taken on 27 February, 6 March and 13 March, respectively, and results are shown in Figure 3.7. On 27 February, it can be observed that clouds were appearing in gap of some time. So, results confirm that when there were clouds in the sky, both tracked panel and static panel generated approximately equal amount of energy. While on 6 March and 13 March, clouds appeared in sky after mid-day. The results confirm that when clouds appeared in sky, both the panels generated equal amount of energy. It was noticed that the behavior of power curve with respect to time was unpredictable due to the presence of clouds.

3.3.2 Clear sky days

To determine the actual performance enhancement using the vertical single-axis solar tracker, readings were taken during clean sky days and the power curves for 28 February, 14 March, 27 March and 3 April are shown in Figure 3.8. Power curves were drawn using readings and best fit polynomial equation for data points were obtained on excel with regression factor of 0.99. For tracked panel data panel, four-degree polynomial was obtained, while for static panel data points, a two-degree polynomial was obtained.

To determine the percentage, enhancement of power generation of tracked panel was determined by using the following equation:

$$\Delta\,(\%) = \frac{E_t - E_s}{E_s} \times 100$$

where
$\Delta\,(\%)$ = Additional power due to vertical axis tracking with respect to static
E_t = Energy generated by tracked panel in a day time
E_s = Energy generated by static panel in a day

In Table 3.3, best fit polynomial equations for tracked and static panel are given with energy generated per day by tracked and static panel and per-centage of additional power due to vertical axis tracking.

Maximum solar radiation around 1000 W/m² was obtained during clean sky days. It was noted that during afternoon hours, there was approximately equal power generation of both tracked panel and static panel. Maximum additional power using the vertical axis tracker was obtained during morning hours when static panel was generating very low power. Power generation of tracked panel during morning and evening session contributed 95% of total additional power.

Polynomial patterns were observed for tracked panel power output and static panel output. For static panel, power at morning and evening session

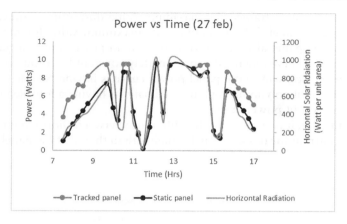

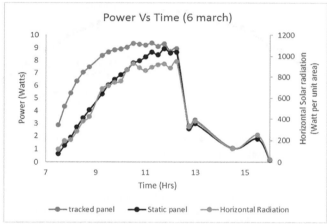

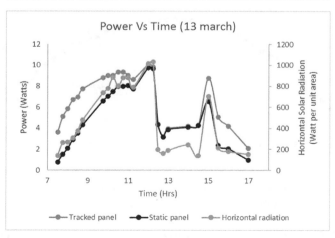

Figure 3.7 Power curve for cloudy sky days.

was quite low and attained the maximum value during noon time. Power curve of static panel did not maintain its maximum value for more time. While tracked panel also achieved its maximum power value during noon time, its power curve remained approximately constant for about five hours. This is the reason why static panel power curve is having two-degree polynomial as best fit polynomial and tracked panel is having a four-degree polynomial for best fit power curve.

In Figure 3.9, it is shown that the difference between short circuit current and peak current for tracked panel is higher than that of static panel, ideally

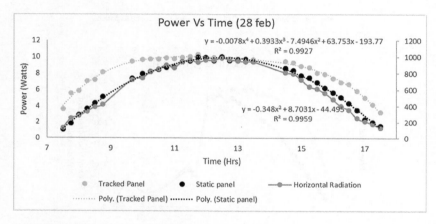

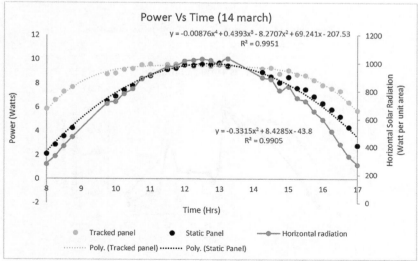

Figure 3.8 Power curve for clear sky days.   (*Continued*)

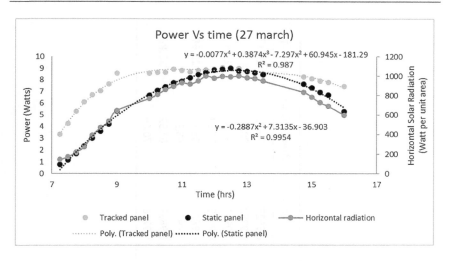

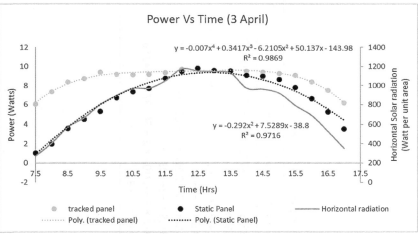

Figure 3.8 (Continued)

which should remain low in order to get high fill factor. So, it will result in better fill factor value of static panel as compared with tracked panel. This is because tracked panel receives more sunlight, which in turn produces more light-generated current, hence high short circuit current. With increase in light-generated current, peak current also increases but not up to the extent of short circuit current. Hence, the difference of short circuit current and peak current in tracked panel remains higher.

Table 3.3 Comparison of performance of tracked panel and static panel

Date	System	Polynomial expression (best fit)	Regression factor (R^2)	Power (Watts)	Δ(%) (Eq. 3.3)
28 Feb	Tracked	y = −0.0078 × 4 + 0.3933 × 3 − 7.4946 × 2 + 63.753 × − 193.77	0.9927	86.13	22.17
	Static	y = −0.348 × 2 + 8.7031 × −44.495	0.9959	70.5	
14 March	Tracked	y = −0.00876 × 4 + 0.4393 × 3 − 8.2707 × 2 + 69.241 × −207.53	0.9951	87.4	23.3
	Static	y = −0.3315 × 2 + 8.4285 × −43.8	0.9905	70.84	
27 March	Tracked	y = −0.0077 × 4 + 0.3874 × 3 − 7.297 × 2 + 60.945 × −181.29	0.987	85.1	23.69
	Static	y = −0.2887 × 2 + 7.3135 × −36.903	0.9954	68.8	
3 April	Tracked	y = −0.007 × 4 + 0.3417 × 3 −6.2105 × 2 + 50.137 × −143.98	0.9869	87.52	23.86
	Static	y = −0.292 × 2 + 7.5289 × −38.8	0.9716	70.66	

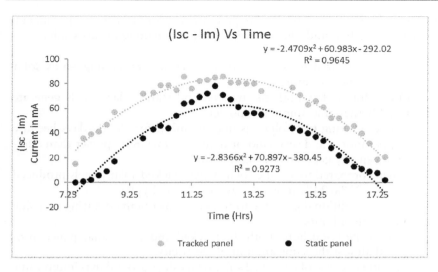

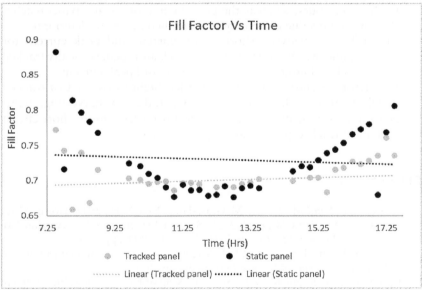

Figure 3.9 (Isc–Im) and fill factor vs Time curve for tracked and static panels on clear sky day.

3.4 CONCLUSION

Solar energy is gaining traction around the world, particularly in India. India has already set some lofty goals for itself to achieve in the future. Solar trackers can play a critical part in the development of the most promising solar power industry in this scenario. The energy production of static solar panels and vertically moved solar panels is explored in this experimental study in Himachal Pradesh. Following are the outcomes of this study:

1. When compared to a static solar panel, a vertical single-axis solar tracker generates more energy.
2. On a cloudy day, both the tracked and static solar panels generate roughly the same amount of power.
3. Active tracking technology is uneconomical for cloudy days because stepper motor and servo motor used consume more power than additional energy generated using tracking.
4. When compared to a static solar panel, tracked solar panels produced average 22%–24% more energy on a clear sky day.
5. 95% of additional gain due to tracking was obtained during morning and evening hours.
6. During the noon hour, both the tracked and static solar panels produced almost the same amount of electricity.
7. The power curve of a tracked solar panel stays around maximum value for five–six hours, whereas the power curve of a static panel reaches the maximum value in noon hours and then begins to deteriorate.
8. The difference between short circuit current and peak current for tracked panel was higher than static panel; although, vertically tracked panel produced large open circuit current and peak current.
9. Fill factor for static panel was recorded higher than that of tracked panel because with increase in solar radiation harvesting, light-generated current increases, which in turn affects more short circuit current of tracked panel than its peak current.

REFERENCES

1. White, F.M., *Fluid Mechanics*, McGraw-Hills, New York, 2011. Green, M.A., Hishikawa, Y., Warta, W., et al. 2020. Solar cell efficiency tables (version 55). *Progressive Photovoltaic Research Applications*, 28(1), pp. 3–15.
2. Kodati, R.B. and Rao, N.P., 2020. A review of solar cell fundamentals and technologies. *Advanced Science Letters*, 26, pp. 260–271.
3. Racharla, S. and Rajan, K., 2017. Solar tracking system: A review. *International Journal of Sustainable Engineering*, 10(2), pp. 72–81.
4. Hafez, A., Yousef, A. and Harag, N., 2018. Solar tracking systems: Technologies and trackers drive types – A review. *Renewable and Sustainable Energy Reviews*, 91, pp. 754–782.
5. AL-Rousan, N., Isa, N. and Desa, M., 2018. Advances in solar photovoltaic tracking systems: A review. *Renewable and Sustainable Energy Reviews*, 82, pp. 2548–2569.
6. Mahendran, M., Ong, H.L., Lee, G.C. and Thanikaikumaran, K., An experimental comparison study between single-axis tracking and fixed photovoltaic solar panel efficiency and power output: Case Study in East Coast Malaysia. *Energy Conversion and Management*, 42, pp. 1205–1214.

7. Bahrami, A., Okoye, C. and Atikol, U., 2016. The effect of latitude on the performance of different solar trackers in Europe and Africa. *Applied Energy*, 177, pp. 896–906.

8. Quesada, G., Guillon, L., Rousse, D., Mehrtash, M., Dutil, Y. and Paradis, P., 2015. Tracking strategy for photovoltaic solar systems in high latitudes. *Energy Conversion and Management*, 103, pp. 147–156.

9. Li, Z., Liu, X. and Tang, R., 2011. Optical performance of vertical single-axis tracked solar panels. *Renewable Energy*, 36(1), pp. 64–68.

10. Bahrami, A. and Okoye, C., 2018. The performance and ranking pattern of PV systems incorporated with solar trackers in the northern hemisphere. *Renewable and Sustainable Energy Reviews*, 97, pp. 138–151.

11. Rahayani, R. and Gunawan, A., 2019. A design and implementation of dual axis solar tracker using mini photo voltaic as solar sensor. *Proceedings of the 8th International Conference on Informatics, Environment, Energy and Applications – IEEA '19*.

12. Wang, J., Zhang, J., Cui, Y. and Bi, X., 2016. Design and implementation of PLC-based automatic sun tracking system for parabolic trough solar concentrator. *MATEC Web of Conferences*, 77, p. 06006.

13. Zhu, Y., Liu, J. and Yang, X., 2020. Design and performance analysis of a solar tracking system with a novel single-axis tracking structure to maximize energy collection. *Applied Energy*, 264, p. 114647.

14. Solanki, C.S. 2019. *Solar Photovoltaics: Fundamentals, Technologies and Applications*. PHI Learning Private Limited, Delhi.

15. El Kadmiri, Z., El Kadmiri, O., Masmoudi, L. and Bargach, M., 2015. A novel solar tracker based on omnidirectional computer vision. *Journal of Solar Energy*, 2015, pp. 1–6.

16. Ray, S. and Tripathi, A., 2016. Design and development of tilted single axis and Azimuth-altitude dual axis solar tracking systems. *2016 IEEE 1st International Conference on Power Electronics, Intelligent Control and Energy Systems (ICPEICES)*.

17. Kaur, T., Mahajan, S., Verma, S., Priyanka and Gambhir, J., 2016. Arduino based low-cost active dual axis solar tracker. *2016 IEEE 1st International Conference on Power Electronics, Intelligent Control and Energy Systems (ICPEICES)*.

Chapter 4

Analysis of drag force on a submerged thin plate under various wave steepness conditions

Numerical simulation approach

Surendra Singh Yadav

National Institute of Technology, Silchar, India

CONTENTS

4.1 INTRODUCTION: BACKGROUND AND DRIVING FORCES

An attempt is made to design a WEC. It is essential to study the wave force and wave behavior in the vicinity of a verticality submerged thin rectangular plate. Numerical techniques are used here to understand the wave–body interaction in the flow field. Cost-efficient and relatively fast initial studies are now feasible with the development of CFD applications. Several authors studied their research problem by numerical method to analyze the wave body problem. Some researchers conducted studies numerically and experimentally on cylinders, rectangular plates, ellipse plates, etc. Roy and Ghosh [1] calculated force on the 2D plate at various submerged depths in shallow water using the Morison equation. Malavasi and Guadagnini [2] examined the free surface effect on rectangular cylinders placed along the flow direction. Arslan et al. [3] observed the significant eddy dynamics effect on a partially submerged rectangular cylinder in the flow field and noted the drag force. Rajagopalan and Niehous [4] conducted a numerical study on a submerged 2D vertical plate to predict the fluid dynamic behavior using second-order stokes wave theory in intermediate water depth. The effect of aspect ratio (AR) on elliptical plates was experimentally studied by Fernando and Rival [5]. Satheesh and Huera-Huarte [6] experimentally examined the drag force of AR at a submergence depth. The drag force increased suddenly due to the recirculating flow in the vicinity of the vertical rectangular plate. Hemmati et al. [7] numerically studied the wave characteristics around 2D

DOI: 10.1201/9781003257691-4

flat plates and calculated lift and drag coefficients. Lui et al. [8] analyzed the free surface effect on 2D vertically rectangular plates under different submerged depths. Finnegan and Goggins [9] studied the wave structure interaction with linear deep-water waves using ANSYS CFX software. Kim et al. [10] numerically investigated wave force on the fixed offshore structure using commercial software (FLUENT). Prasad et al. [11] modeled a 3D numerical wave tank (NWT) to simulate wave characteristics using the ANSYS CFX software. The authors also validated numerical results with experimental results. X. Tian et al. [12] modeled 3D NWT to analyze the wave load on the structure using Stoke's second-order wave theory. The present study numerically investigated drag force on a submerged vertical plate in two different cases (Type 1 and Type 2) under various wave steepness conditions. Second-order stokes wave theory is used in this study in intermediate water depth to analyze the problem. In this study, numerical investigations were done by ANSYS FLUENT commercial software in a 2D NWT.

4.2 METHODOLOGY

The two-dimensional 2-D NWT is constructed as a rectangular domain $(l_x \times l_z)$ with various boundary conditions, as shown in Figure 4.1. The inflow method is applied for wave generation at the inlet boundary. The top boundary is employed as a pressure outlet that is considered open to the atmosphere. The bottom and right boundaries of the NWT are specified as solid surfaces with the no-slip wall constraint applied. Due to the no-slip boundary condition, the velocity of fluid near the solid wall is zero. The x-axis is measured positively in the direction of wave propagation, while the z-axis is measured vertically from the still water level (SWL) upward. The influence of y-direction is not taken due to consideration of the 2-D domain. The computational domain is divided into three sections: a) wave generating zone (l_{x1} = 2 m); b) working zone (l_{x2} = 6 m); and c) damping zone (l_{x3} = 2 m). The overall size of the NWT is l_x = 10 m and l_z = 1 m. In this study, the inflow method generates second-order stokes wave of time period T = 1.1658 sec and wavelength L = 1.8482 m. The structure grid has been utilized to discretize the computational domain. The mesh refinement has been implemented to capture the free surface in the fluid domain accurately.

Implementing a sloping beach as a damping zone in the NWT is important for preventing the wave reflection from the end wall; otherwise, waves reflect from the end of the tank and the simulation disorders. The beach of 1:5 slope has been formed in the damping zone. Figure 4.2 represents the uniformly structured grid mesh configuration of 250,000 nodes. Total work has been simulated with a system specification of Intel(R) Xeon(R) W-2155CPU@3.30 GHz 3.31 GHz processor and a 64.0 GB RAM. The

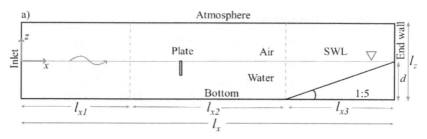

Type 1: surface –piercing plate

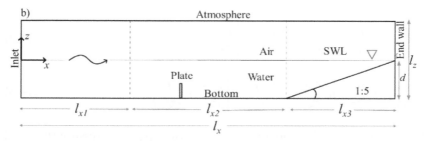

Type 2: bottom standing plate

Figure 4.1 Schematic of 2D NWT for two types of plate positions.

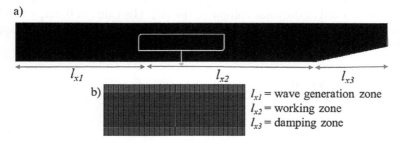

Figure 4.2 Computational mesh in the numerical domain.

simulation has been carried out with 5000 time steps, a time-step size of 0.01 sec, and the maximum iteration has been selected as 20. The Courant number is set at 0.25. Table 4.1 shows the wave parameters utilized in this investigation.

Water wave propagation in a NWT is modeled with commercial ANSYS FLUENT software based on FVM. The flow behavior is assumed to be irrotational, without surface tension, and with atmospheric pressure $p_a = 0$. The mass continuity and Navier–Stokes equations (NSE) are used as governing equations of an unsteady, incompressible flow of the present problem are described.

Table 4.1 Wave parameters

	d(m)	L(m)	T(sec)	H(m)	(d/L)	(H/L)
1	0.4	1.8482	1.1658	0.05	0.2164	0.0271
2	0.4	1.8482	1.1658	0.06	0.2164	0.0325
3	0.4	1.8482	1.1658	0.07	0.2164	0.0379

$$\frac{\partial u}{\partial x} + \frac{\partial w}{\partial z} = 0 \tag{4.1}$$

$$\frac{\partial u}{\partial t} + u\frac{\partial u}{\partial x} + w\frac{\partial u}{\partial z} = -\frac{1}{\rho}\frac{\partial p}{\partial x} + v\left[\frac{\partial^2 u}{\partial x^2} + \frac{\partial^2 u}{\partial z^2}\right] \tag{4.2}$$

$$\frac{\partial w}{\partial t} + u\frac{\partial w}{\partial x} + w\frac{\partial w}{\partial z} = -\frac{1}{\rho}\frac{\partial p}{\partial z} + v\left[\frac{\partial^2 w}{\partial x^2} + \frac{\partial^2 w}{\partial z^2}\right] \tag{4.3}$$

where u and w are the velocity [m s^{-1}], t = time [s], p = pressure [Pa], and ρ = density [kg m^{-3}] which varies over the air and water interface, and v is the kinetic viscosity [m^2 s^{-1}]. In this analysis, the interface of two fluids is modeled based on the VOF method (Hirt and Nichols [13]). The volume fraction function α_q is calculated as the ratio of the volume occupied by the qth phase in a cell to the total volume of the cell. The value of $0 < \alpha_q < 1$ indicates that the cell is an interface cell. The following equations are employed to determine α_q:

$$\frac{\partial \alpha_q}{\partial t} + \nabla \cdot (\alpha_q \bar{V}) = 0 \tag{4.4}$$

$$\sum_{q=1}^{2} \alpha_q = 1 \tag{4.5}$$

where $\bar{V}$ = velocity vector. The two-phase mixture density is calculated as follows:

$$\rho = \alpha_q \rho_w + (1 - \alpha_q)\rho_a \tag{4.6}$$

where ρ_a = 1.225 kg/m^3 and ρ_w= 998.2 kg/m^3 are the air and water density, respectively.

Equation (1.7) represents the surface elevation of wave as follows:

$$\eta(x,t) = \frac{H}{2}\cdot\cos(kx-\omega t) + \frac{kH^2}{16}\cdot\frac{\cosh(kd)}{\sinh^3(kd)}\cdot(2+\cosh 2kd)\cdot\cos 2(kx-wt) \tag{4.7}$$

where the angular frequency of wave $\omega = 2\pi/T$ [rad s^{-1}]; wave number $k = 2\pi/L$; x is the longitudinal direction [m] along the NWT length.

The drag coefficient is calculated as follows:

$$C_d = \frac{F_x}{\frac{1}{2}\rho A u^2} \qquad (4.8)$$

where C_d is the drag coefficient, F_x is the horizontal wave force [N]; A is the projected area of thin plate [m^2].

4.3 RESULTS AND DISCUSSION

The present numerical study investigated the analysis of drag forces of ocean waves on a thin rectangular plate under two cases (Type 1 and Type 2) at $d = 0.40$ m and $T = 1.1658$ sec. The parameters of wave used in this study are as shown in Table 4.1. Figure 4.3 depicts a plot of the horizontal wave force on the plate vs time for two different geometrical cases (Type 1 and Type 2) and three wave steepness conditions. The graph is showing periodic.

Initially, wave force on the plate increases to a maximum value between the time 38.32 sec and 38.65 sec and then decreases from positive to negative. Figure 4.3 (a, b) shows that the wave force on the plate is higher for wave steepness $H/L = 0.0379$ than $H/L = 0.0271$ and 0.0325. It has also been observed that wave forces on the plate for Type 1 experienced more than Type 2.

Figure 4.4 illustrates the wave force (F_x) for Type 1 and Type 2 against H/L at $t = 39.65$ sec and $d = 0.4$ m. It has been shown from Figure 4.4 that forces increase with the rise of wave steepness. It has also been shown from

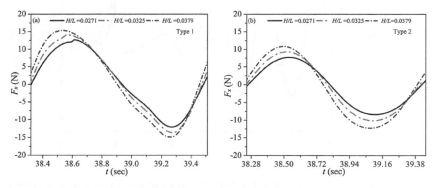

Figure 4.3 Wave force (F_x) versus time (t) at $d = 0.40$ m and $T = 1.1658$ sec for the Type 1 and Type 2.

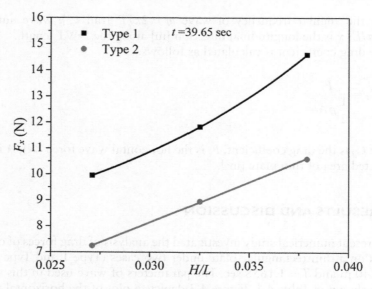

Figure 4.4 Wave force (F_x) versus wave steepness (H/L) at d = 0.40 m and
T = 1.1658 sec for the Type 1 and Type 2.

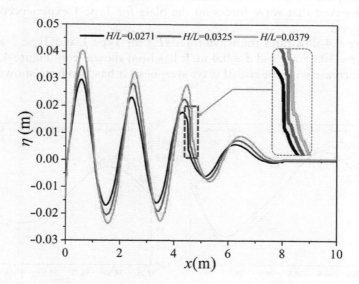

Figure 4.5 Comparison of surface elevation (η) near the vicinity of the plate at
various wave steepness conditions.

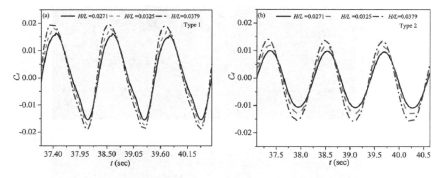

Figure 4.6 Comparison of drag coefficients (C_d) for the Type I and Type 2 versus time (t) under various H/L.

Figure 4.4 that the wave force for Type 1 is more as compared with Type 2. The reason behind this water particle velocity on $z = 0$ m (SWL) is more, and the velocity rate gradually decreases toward the bottom and finally becomes zero.

Figure 4.5 illustrates the comparison of surface elevation (η) near the vicinity of plate at various wave steepness conditions. The zoom area shows the accurate observation of the lines. It has been shown in Figure 4.5 that the nonlinearity effect developed more at higher wave steepness ($H/L = 0.0379$) conditions, and at this wave steepness vortex strength is more than other wave steepness conditions ($H/L = 0.0271$ and 0.0325).

The drag coefficient is shown as a function of time at two distinct positions (Type 1 and Type 2) of the plate, as illustrated in Figure 4.6 for various H/L values. The drag coefficient has been calculated at a flow time of 37.15 sec –40.65 sec to avoid the initial transient effects. The trends of drag coefficients are periodic, but the magnitudes are different. The drag coefficient value has shown a higher value for the wave steepness ($H/L = 0.0379$) in both cases implying that the surging force on the plate at this wave steepness is maximum.

Figure 4.7 shows the plot of the velocity vector and vorticity fields around the plate at the surface-piercing (Type 1) position under different wave steepness conditions at $d = 0.40$ m and $T = 1.1658$ sec. Figure 4.7 (a) demonstrated flow field for the wave steepness $H/L = 0.0271$ at time $t = 38.62$ sec and Figure 4.7 (b) demonstrated flow field for the $H/L = 0.0379$ at time $t = 38.5$ sec. It has been observed from Figure 4.7(b) that the strength of vortices is more as compared with Figure 4.7(a). The reason behind this is that Figure 4.7(b) experiences more wave force at time $t = 38.5$ sec than Figure 4.7(a). As the plate is vertically submerged, the upstream velocity vectors travel on the plate in the propagation direction, and downstream velocity vectors travel upward. Figure 4.7(a, b) also shows that velocity vectors on the plate zero due to the no-slip condition.

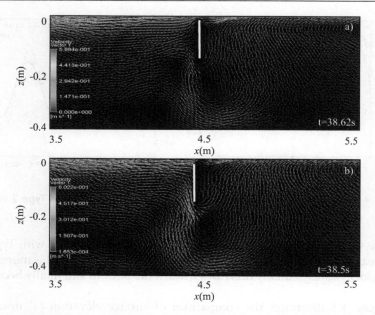

Figure 4.7 Comparison of flow field distribution around the plate under various *H/L*.

4.4 CONCLUSIONS

The present model investigated drag force using second-order stokes wave theory in the NWT under two different cases (Type 1 and Type 2). The model has been developed based on CFD code ANSYS FLUENT software to achieve this target. The VOF method is used to track the interface of fluids. The inflow velocity method and the Dirichlet boundary condition are used to produce the desired nonlinear waves in the NWT. Velocity vector and surface profile are analyzed in the velocity field. Our findings support the following conclusions based on the above numerical models.

- Horizontal wave force (F_x) increases with the wave steepness (H/L). The maximum force of the system is reached at $H/L = 0.0379$.
- Horizontal wave force (F_x) for Type 1 is more than Type 2.
- Horizontal wave force (F_x) versus time (t) curve periodically in nature.
- Drag coefficient (C_d) periodic with respect to time (t), and it is higher for H/L (wave steepness) = 0.0379, implying that F_x on the plate is maximum at the wave steepness.
- Drag coefficient (C_d) for Type 1 is more compared to Type 2.

- Vortex strength of Figure 4.7(b) is more as compared with Figure 4.7(a). The reason behind this is that Figure 4.7(b) experiences more wave force at time $t = 38.5$ sec than Figure 4.7(a).
- Figure 4.7(a, b) also illustrates that velocity vectors on the plate are zero due to the no-slip condition.

REFERENCES

1. P. D. Roy and S. Gosh, "Wave force on vertically submerged circular thin plate in shallow water." *Ocean Engineering*, vol. 33, pp. 1935–1953, 2006. https://doi.org/10.1016/j.oceaneng.2005.09.010
2. S. Malavasi and A. Guadagnini, "Interactions between a rectangular cylinder and a free-surface flow." *Journal of Fluids and Structures*, vol. 23, pp. 1137–1148, 2007. https://doi.org/10.1016/j.jfluidstructs.2007.04.002
3. T. Arslan, S. Malavasi, B. Pettersen and H. I. Andersson, "Turbulent flow around a semi-submerged rectangular cylinder." *Journal of Offshore Mechanics and Arctic Engineering*, vol. 135, pp. 1–11, 2013. https://doi.org/10.1115/1.4025144
4. K. Rajagopalan and G. Nihous, "Study of the force coefficients on plates using an open source numerical wave tank." *Ocean Engineering*, vol. 118, pp. 187–203, 2016. https://doi.org/10.1016/j.oceaneng.2016.03.028
5. J. N. Fernando and D. E. Rival, "Reynolds-number scaling of vortex pinch-off on low aspect-ratio propulsors." *Journal of Fluid Mechanics*, vol. 799, 2016bR3. https://doi.org/10.1017/jfm.2016.396
6. S. Satheesh and F. J. Huera-Huarte, "Effect of free surface on a flat plate translating normal to the flow." *Ocean Engineering*, vol. 171, pp. 458–468, 2019. https://doi.org/10.1016/j.oceaneng.2018.11.015
7. A. Hemmati, D. H. Wood and R. J. Martinuzzi, "Characteristics of distinct flow regimes in the wake of an infinite span normal thin flat plate." *International Journal of Heat and Fluid Flow*, vol. 62, pp. 423–436, 2016. https://doi.org/10.1016/j.ijheatfluidflow.2016.09.001
8. I. H. Liu, J. Riglin, W. C. Schleicher and A. Oztekin, "Flow past a plate in the vicinity of a free surface." *Ocean Engineering*, vol. 111, pp. 323–334 2016. https://doi.org/10.1016/j.oceaneng.2015.11.009
9. W. Finnegan and J. Goggins, "Numerical simulation of linear water waves and wave structure interaction." *Ocean Engineering*, vol. 43, pp. 23–31, 2012. https://doi.org/10.1016/j.oceaneng.2012.01.002
10. S. Y. Kim, K. M. Kim, J. C. Park, G. M. Jeon and H. H. Chun, "Numerical simulation of wave and current interaction with a fixed offshore substructure." *International Journal of Naval Architecture and Ocean Engineering*, vol. 8, pp. 188–197, 2016. https://doi.org/10.1016/j.ijnaoe.2016.02.002
11. D. D. Prasad, M. R. Ahmed, Y. H. Lee and R. N. Sharma, "Validation of a piston type wave-maker using numerical wave tank." *Ocean Engineering*, vol. 131, pp. 57–67, 2017. https://doi.org/10.1016/j.oceaneng.2016.12.031

12. X. Tian, Q. Wang, G. Liu, W. Deng and Z. Gao, "Numerical and experimental studies on a three-dimensional numerical wave tank." *IEEE Access*, vol. 6, pp. 6585–6593, 2018. https://doi.org/10.1109/ACCESS.2018.2794064

13. C. W. Hirt and B. D. Nichols (1981), "Volume of Fluid (VOF) method for the dynamics of free boundaries." *Journal of Computational Physics*, vol. 39, pp. 210–225, 1981. https://doi.org/10.1016/0021-9991(81)90145-5

Chapter 5

A single-zone thermodynamic model cycle simulation and optimization of combustion and performance characteristics of a diesel engine

Puneet Singh Gautam, Pradeep Kumar Vishnoi and V. K. Gupta

G. B. Pant University of Agriculture & Technology, Pantnagar, India

CONTENTS

5.1 INTRODUCTION

Transportation is the widespread medium used by people not only to communicate with one another but also to advance. The economical, environmentally friendly transportation system requires the development of novel automotive technologies which involve technologies used in the design and prototype, which has become critical. These technologies are necessary for existing vehicles to evolve and adapt. On the other hand, modeling in science and technology can represent physical phenomena in a given system using mathematical equations (with reasonable assumptions) [1, 2]. Therefore, the modeling of engine combustion processes is of great advantage. Engine combustion modeling can be categorized into "zero-dimensional, quasi-dimensional, and multi-dimensional models." The difficulty and complexity

DOI: 10.1201/9781003257691-5

of constructing mathematical equations and their computational time increase for higher-dimensional models.

On the other hand, the computed results are closer to the practical results than the zero-dimensional model. In zero-dimensional models, there are no flow field dimensions. All of the procedures occur instantaneously [3]. Despite the shortcomings of quasi-dimensional and multi-dimensional models, a zero-dimensional model is chosen because of its simplicity and ease of use in predicting the numerical results for a diesel engine in terms of combustion (HRR, CP, ROPR, mean gas temperature, and performance characteristics (BP, BSFC, and BTE).

5.2 LITERATURE REVIEW AND OBJECTIVE

Nabi et al. [4] investigated a thermodynamic analysis of the engine's performance and combustion characteristics by developing a one-dimensional model incorporating various sub-models for combustion and HRR. Gogoi [5] and Baruah investigated thermodynamic cycle simulation analysis to predict the performance parameters of a compression ignition engine fueled with biodiesel-diesel blends. Awad et al. [6] investigated a diesel engine fueled with biodiesel of waste cooking oil and animal fat wastes and developed a thermodynamic model to validate cylinder pressure with the model. Hariram and Bharathwaaj [7] experimented on a one-cylinder four-stroke diesel engine running on neat diesel, and beeswax biodiesel blends with the help of a zero-dimensional model and predicted the cylinder pressure, ROHR, and ROPR and found that the predicted results were in good agreement with experimental values. Based on a thorough review of the published literature, there are no articles published on a zero-dimensional thermodynamic model on a compression ignition engine for the evaluation of combustion parameters such as CP, ROPR, HRR, ID, etc. and engine performance characteristics such as indicated power (IP) and torque (IT), brake thermal efficiency (BTE), frictional torque, mean effective pressure (FT, FP & FMEP), etc. with more variables at a time.

This study explored the combustion and performance characteristics of a one-cylinder four-stroke diesel engine using a thermodynamic single-zone zero-dimensional simulation model. The thermodynamic model was developed by coding the script in MATLAB software due to its simplicity and less computational time. The engine speed varying from 900 to 2400 rpm, compression ratio from 12 to 24, injection timing from 24° crank angle (CA) bTDC to 18° crank angle aTDC, and exhaust gas recirculation (EGR) from 10% to 90% were chosen as engine's variables against which the combustion and performance parameters were evaluated. The impact of engine speeds, compression ratios, injection timings, and EGR on engine combustion and performance characteristics was explored in this study, and the findings were addressed. The optimization of the performance parameters was also done using response surface methodology (RSM) based on BMEP, BTE, and BT.

5.3 MATERIALS AND METHODS

A one-cylinder four-stroke diesel engine with a constant 1500 rpm, a bore/stroke of 80/110 mm, and a compression ratio of 16.5 with an injection timing of −24° bTDC was selected for the modeling.

5.3.1 Mathematical modeling of engine combustion

During the compression and expansion strokes of a diesel engine, thermodynamic analysis was performed to develop a mathematical zero-dimensional model for the theoretical analysis of engine combustion characteristics and engine performance characteristics fueled with regular diesel fuel. Vibe's correlation [8] shows the total chemical energy released by the fuel as a function of crank angle. The form parameter (m) and the BMF curve's point of inflexion location determine the rate of pressure rise.

$$\chi(\theta) = 1 - \exp\left[-a\left(\frac{\theta-\theta_i}{\theta_d}\right)^{(m+1)}\right] \tag{5.1}$$

where "m" and "a" are the parameters that can be computed by trial and error for the engine [8].

The HRR can be expressed in terms of η_c (combustion efficiency), Vibe's correlation and LHV (the lower heating value).

$$\frac{dQ}{d\theta} = \frac{dQ_{in}}{d\theta} - \frac{dQ_{loss}}{d\theta} \tag{5.2}$$

$$\frac{dQ}{d\theta} = \left(\eta_c \times m_f \times LHV \times \frac{d\chi}{d\theta}\right) - \frac{dQ_{loss}}{d\theta} \tag{5.3}$$

The Equation (5.4) can be used to compute in-cylinder pressure

$$\frac{dP}{d\theta} = \frac{Y-1}{V}\left(\frac{dQ}{d\theta}\right) - Y\frac{P}{V}\left(\frac{dV}{d\theta}\right) \tag{5.4}$$

The Equation (5.5) can be used to evaluate the volume at each crank angle (θ) [1, 3].

$$V(\theta) = V_c + \frac{\pi B^2}{4}[l + a - s(\theta)] \tag{5.5}$$

where V_c = clearance volume and s are calculated using the following equation:

$$s(\theta) = a\cos(\theta) + (l^2 - a2\sin(\theta^2)^{\frac{1}{2}}$$

The in-cylinder temperature can be found out by employing the first law of thermodynamics which is expressed as:

$$\frac{dT}{d\theta} = \frac{T}{PV}(\gamma - 1)\frac{dQ}{d\theta} - \frac{T}{V}(\gamma - 1)\frac{dV}{d\theta} \qquad (5.6)$$

The heat loss from the burning of fuel to the cylinder wall is computed by the correlation given in Equation 5.7 for heat transfer [2]:

$$\frac{dQ_{loss}}{d\theta} = (h_c(\theta) + h_r(\theta)) \cdot A(\theta) \cdot (T_\theta - T_w) \cdot \left(\frac{1}{\omega}\right) \qquad (5.7)$$

where

$$h_c(\theta) = \frac{k_g Nu}{B} \qquad (5.8)$$

$$h_r(\theta) = 4.25 \times 10^{-9}\left(\frac{T_\theta^4 - T_w^4}{T_\theta - T_w}\right) \qquad (5.9)$$

where ω = woschni's factor, T_w = cylinder wall temperature.

The FMEP for a CI engine may be calculated as suggested by Heywood [1]

$$FMEP = C + 48 \times \left(\frac{RPM}{1000}\right) + 0.4 \times U_p^2 \qquad (5.10)$$

The mechanical work delivered during the piston's compression and expansion stroke process is total or gross indicated work. The following formulae can compute indicated work and indicated mean effective pressure (IMEP) [9].

$$W_{ind} = \int_{-180}^{180} \left(P(\theta)\frac{dV}{d\theta}\right)d\theta \qquad (5.11)$$

$$\text{IMEP} = \frac{W_{\text{ind}}}{V_s} \tag{5.12}$$

Equations (5.13) and (5.14) computed the ITE and ISFC.

$$\text{ITE} = \frac{IP}{m_f \times LHV} \tag{5.13}$$

$$\text{ISFC} = \frac{m_f}{IP} \tag{5.14}$$

Similarly, BMEP, BSFC, and BTE are calculated using BP and BT in the above equations. The model was developed for a one-cylinder, four-stroke diesel engine having a bore and stroke of 80 and 110 mm, a compression ratio of 16.5, and an injection timing of −24° bTDC. For thermodynamic analysis of engine, the model required input parameters such as bore, stroke, connecting rod length, atmospheric temperature, and pressure, intake, and exhaust valve opening/closing timing, etc. to find out the combustion parameters of the engine such as CP, ROPR, HRR, and in-cylinder temperature, etc.

5.4 RESULTS AND DISCUSSION

5.4.1 Injection timing's impact on engine performance

The impact of injection timing on various engine performance parameters is shown in Figure 5.1. The indicated brake and frictional torque are depicted in Figure 5.1a, whereas engine powers and related mean effective pressures are shown in Figure 5.1b and Figure 5.1c, respectively. Figure 5.1d illustrates fuel consumption and associated thermal efficiencies. The engine conditions were set at an engine speed of 1500 rpm, EGR at 0%, a compression ratio of 16.5, and a full load position. Figure 5.1a shows the indicated brake and frictional torque variation versus injection timing. The indicated torque and brake torque increase to a maximum at about −10° bTDC, and then decrease afterward. This decrease is due to the combustion chamber's ignition delay. The frictional term is a function of mean piston speed and engine speed, as shown in equation (5.10); therefore, there is no effect on frictional torque of injection timing. Figure 5.1b presents the engine powers such as indicated, brake, and frictional power under the similar operating conditions stated in Figure 5.1a.

As shown in Figure 1a, the engine powers follow the same pattern. Likewise, Figure 5.1c depicts the related mean effective pressures variation

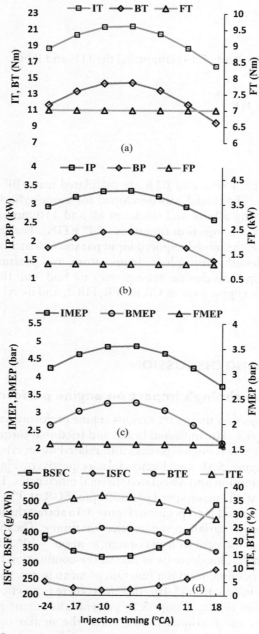

Figure 5.1 Injection timing's impact on (a) Engine torques; (b) Powers; (c) MEPs; (d) Fuel consumption; and thermal efficiencies.

and follows the same trend as Figures 5.1a and 5.1b. Figure 5.1d illustrates the ISFC and BSFC on the left y-axis, whereas ITE and BTE are shown on the right y-axis. The ISFC and BSFC first decrease to a minimum at around −10° bTDC and increase again afterward. This is because, at −10° bTDC, there is a shorter ignition delay in combustion, and more power is created. Equations (5.13) and (5.14) indicate that the ITE and BTE are reciprocals of ISFC and BSFC. Figure 5.1d shows that the best injection timing of −10° bTDC was observed at which the BTE was found maximum. Nabi et al. [4] investigated the injection timing's effect on engine performance and discovered comparable patterns in terms of performance such as ISFC, BSFC, ITE, and BTE. The results obtained are very much comparable to those obtained by Gautam et al. [10, 11], Nabi et al. [4], and Nabi and Rasul [12].

5.4.2 The impact of engine speed on performance

The impact of engine speed is shown in Figure 5.2 on different engine performance characteristics. Figure 5.2a depicts torques (indicated, brake, and frictional torque), while Figure 5.2b and 5.2c show the corresponding power and mean effective pressure (indicated, brake, and frictional). Figure 5.2d illustrates fuel consumption and related thermal efficiencies. The engine conditions were fixed at an injection timing of −24° bTDC, EGR at 0%, a compression ratio of 16.5, and full load position. Figure 5.2a shows the IT, BT, and FT variation versus engine speed. The IT increases slightly and decreases after 2150 rpm, but BT decreases as the engine speed increases. The decrease in BT is due to the rise in FT. The indicated, brake, and frictional power are shown in Figure 5.2b under the identical operating conditions as previously mentioned. Engine powers shown in Figure 5.2c follow the same pattern as engine torques in Figure 5.2a.

Similarly, mean effective pressure variations are shown in Figure 5.2c, which follows the same trend as Figure 5.2a. On the left y-axis is the ISFC, BSFC, and on the right y-axis is the ITE, BTE. With increased engine speed, the ITE increases while the BTE reduces. The BTE is dependent on BP, which depends on the frictional losses dependent on engine speed, as can be seen from equation (5.10), and with increasing engine speed, BSFC rises. Fuel consumption increases as engine speed increases, resulting in a rise in the BSFC. Similar results were also reported by Nabi et al. [4], Vishnoi et al. [13], and Gautam et al. [14].

5.4.3 Compression ratio's impact on engine performance

Figure 5.3 shows the compression ratio's impact on various engine performance characteristics. Figure 5.3a presents torques (indicated, brake, and frictional torque), while Figure 5.3b and 5.3c show the corresponding

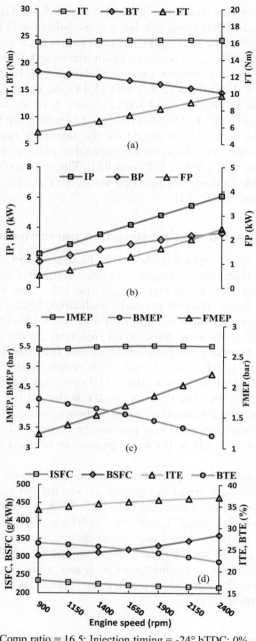

Comp ratio = 16.5; Injection timing = -24° bTDC; 0%
EGR; Full load

Figure 5.2 Impact of engine speed on (a) Engine torques; (b) Powers; (c) MEPs;
(d) Fuel consumptions; and thermal efficiencies.

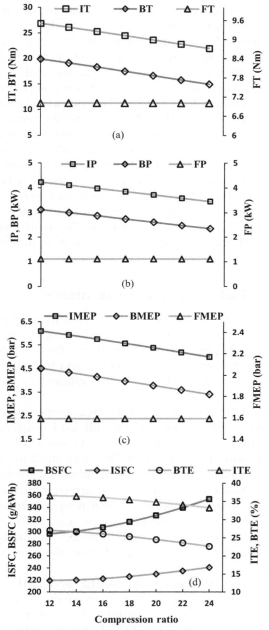

Injection timing = -24° bTDC; Engine speed = 1500 rpm;
0% EGR; Full load

Figure 5.3 Compression ratio's impact on (a) Engine torques; (b) Powers;
(c) MEPs; (d) Fuel consumptions; and thermal efficiencies.

power and mean effective pressure (indicated, brake, and frictional), respectively, and fuel consumptions and related thermal efficiencies are illustrated in Figure 5.3d. The injection timing was set to –24° bTDC, the EGR to 0%, the engine speed to 1500 rpm, and the full load position was used. Figure 5.3a shows the variation of torques (indicated, brake, and frictional torque) versus compression ratio. As the compression ratio rises, the indicated and brake torque reduces. The increase in inlet temperature at BDC is causing the drop in IT and BT. The engine powers are shown in Figure 5.3b under the identical operating parameters as previously mentioned. Indicated, brake, and frictional powers follow the same pattern as the engine torques. Figure 5.3c shows the variations of mean effective pressures and other parameters and follows the same trend as Figure 5.3a. On the left y-axis is the ISFC, BSFC, and on the right y-axis is the ITE, BTE. With a higher compression ratio, the ITE and BTE reduce. The ISFC and BSFC increase as the compression ratio rises. The trends can be validated with the results obtained by Nabi et al. [4] and Nabi and Rasul [12] and Gautam et al. [11].

5.4.4 Effect of EGR on engine performance

Figure 5.4 shows the impact of EGR on various engine performance parameters. Figure 5.4a presents torques (indicated, brake, and frictional torque). In contrast, Figures 5.4b and 5.4c depict the corresponding power and mean effective pressure (indicated, brake, and frictional), respectively, and fuel consumptions and related thermal efficiencies are presented in Figure 5.4d.

The injection timing was set to –24° bTDC, compression ratio to 16.5, the engine speed to 1500 rpm, and the full load position. Figure 5.4a shows the variation of the engine torques versus EGR. The indicated torque and brake torque increase as EGR is enhanced from 0% to 90%. The increase in IT and BT is due to better combustion efficiency of the air-fuel mixture as residuals from the exhaust are recirculated into an inlet manifold where it mixes with air at a higher temperature before going to the combustion chamber. The engine powers are shown in Figure 5.4b under the identical operating conditions as previously mentioned. Indicated, brake, and frictional powers follow the same pattern as the engine torques.

The mean effective pressures IMEP, BMEP, and FMEP are shown in Figure 5.4c, which follows the same pattern as the other parameters in Figure 5.4a. Figure 5.4d illustrates the ISFC, BSFC on the left y-axis, and ITE, BTE on the right y-axis. The ITE decreases slightly, whereas BTE increases to a maximum at around 15% EGR and declines with a further EGR increase. The ISFC increases and BSFC reduces to a minimum at 15% EGR, then increases afterward as it is reciprocal of BTE. The optimum EGR in terms of BTE is found to be at 15% EGR from the figure.

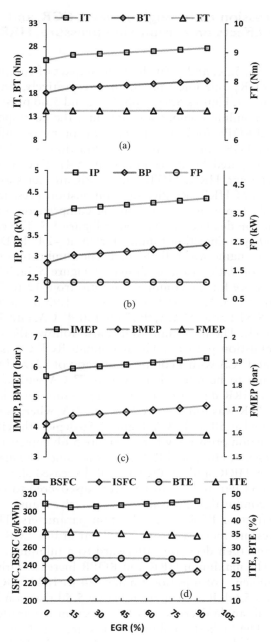

Comp. ratio= 16.5; Injection timing = -24° bTDC;
Speed=1500 rpm; Full load

Figure 5.4 Effect of EGR on (a) Engine torques; (b) Powers; (c) MEPs; (d) Fuel consumptions; and thermal efficiencies.

5.4.5 Compression ratio, engine speed, EGR, and injection timing effects on combustion pressure, HRR, and ROPR

Figures 5.5a, 5.5b, 5.5c, and 5.5d show the impact of compression ratio, engine speed, EGR, and injection timing, respectively, on combustion pressure. In contrast, Figures 5.6a, 5.6b, 5.6c, and 5.6d present maximum values of corresponding combustion pressure, the corresponding crank angle, HRR, and ROPR for different compression ratios, different engine speeds, different EGR, and different injection timings. The combustion pressure was computed by using Equation (5.4). Due to space limitations for this chapter, HRR, MGT, and ROPR are not shown here, but maximum values of HRR, ROPR, and combustion pressure are shown in Figure 5.6. Peak pressure occurs during compression and expansion strokes near the top dead center, as seen in Figure 5.5. The engine's speed was set at 1500 rpm, injection timing started at −24° bTDC, there was no EGR, and the engine was fully loaded. The combustion pressure rises as the compression ratio rises, as shown in Figure 5.5a. The maximum combustion pressure for the compression ratio from 12 to 24 was found to be 55.09 bar at 7° CA, 63.53 bar at 6° CA, 71.93 bar at 6° CA, 80.26 bar at 5° CA, 88.51 bar at 5° CA, 96.73 bar at 4° CA, and 104.94 bar at 4° CA, respectively. The maximum ROPR also follows the same trend as maximum combustion pressure. The maximum ROPR was found to be 7.91 bar/° CA at 4° CA with a compression ratio of 12, and the minimum ROPR was found to be 4.072 bar/° CA at 7° CA with a compression ratio of 24. The HRR decreases as the compression ratio increases, and the maximum HRR was found to be 57.23 J/° CA, whereas the minimum HRR was obtained as 50.56 J/° CA. Figure 5.5b depicts the effect of engine speed on combustion pressure, whereas Figure 5.6b depicts the maximum combustion pressure, HRR, and ROPR for the same. The maximum combustion pressure, HRR, and ROPR have all decreased slightly, as seen in Figure 5.6b. 73.34 bar and 72.56 bar were discovered to be the highest and minimum combustion pressures, respectively. Figure 5.5c and 5.6c illustrate the effect of EGR on combustion parameters. The maximum combustion pressure increases from 73.99 bar at 0% EGR to 74.95 bar at 15% EGR and then decreases afterward to a minimum of 73.91 bar at 90% EGR. The maximum HRR and ROPR increase with the increase in EGR. The influence of injection timing −24 to 18° bTDC is shown in Figures 5.5d and 5.6d. It is clear from Figure 5.5d that combustion pressure versus crank angle decreases as the injection is retarded and hence ROPR as well. The highest combustion pressure was found to be 82.18 bar at 3° CA, and the minimum was found to be 36.36 bar at −1° CA. Gautam et al. [11], Vishnoi et al. [15], and Nabi et al. [4] and Nabi and Rasul [12] found similar results.

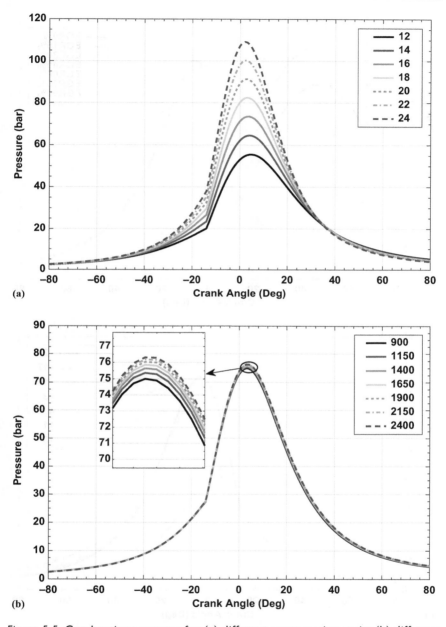

Figure 5.5 Combustion pressure for (a) different compression ratio; (b) different engine speed;

(*Continued*)

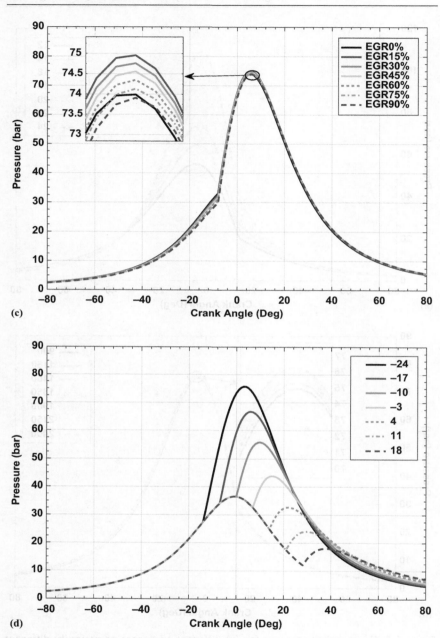

(c)

(d)

Figure 5.5 (Continued) (c) different exhaust gas recirculation (EGR); (d) different injection timing.

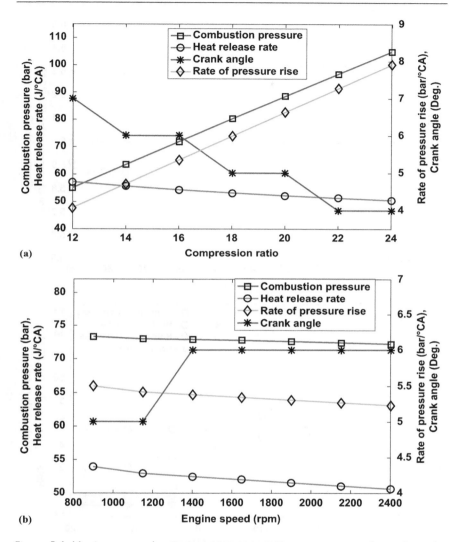

(a)

(b)

Figure 5.6 Maximum combustion pressure, maximum pressure of crank angle, maximum HRR, maximum ROPR for (a) different compression ratios; (b) different engine speeds;

(Continued)

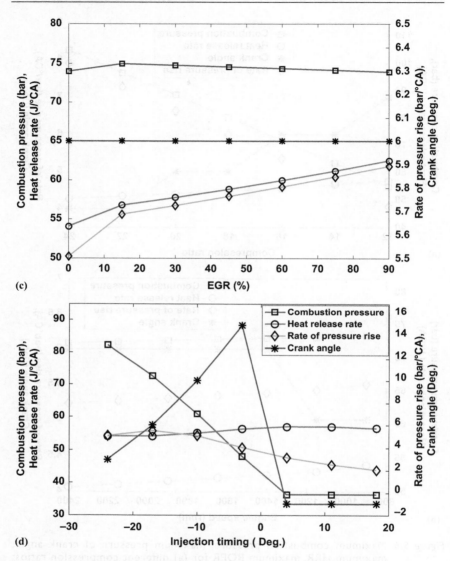

Figure 5.6 (Continued) (c) different EGRs; (d) different injection timings.

Figure 5.7 shows the optimum value of engine variables (engine speed, compression ratio, EGR, and injection timing) using RSM based on BTE, BMEP, and BT. The optimum values were found at an engine speed of 1278 rpm, a compression ratio of 12 and EGR of 80%, and –24° bTDC for the start of injection timing.

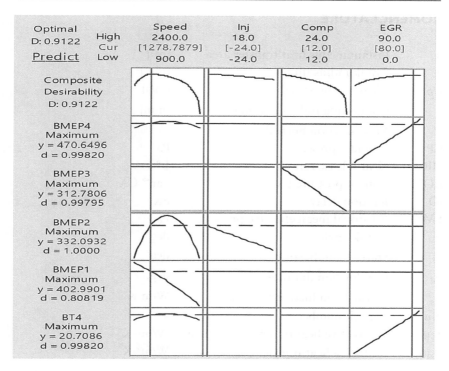

Figure 5.7 A response surface methodology for optimization of engine variables.

5.5 CONCLUSIONS

The investigation was done on prediction of a four-stroke CI engine's combustion and performance characteristics using zero-dimensional thermodynamic model. The theoretical values obtained were in closer agreement with those in the literature cited above. The RSM was applied to find the optimum values of variables such as engine speed, compression ratio, EGR, and injection timing in terms of engine performance parameters such as BTE, BMEP, and BT at 1278 rpm, 12 and EGR of 80%, and –24° bTDC for the start of injection timing.

ACKNOWLEDGMENTS

The author wants to thank G. B. Pant University of Agriculture and Technology and project mentor Dr. V. K. Gupta for his supervision and constructive recommendations during the study.

NOMENCLATURE

A	Instantaneous heat transfer area	[m²]
a	Crank radius (mm)	[mm]
Up	Mean piston speed	[m/s]
l	Connecting rod length (mm)	[mm]
χ	Mass fraction burned	–
CP	Cylinder pressure	Pa
HRR	Heat release rate	J/° CA
ROPR	Rate of pressure rise	bar/° CA
ID	Ignition delay	ms
FMEP	Frictional mean effective pressure	Pa
θ	Crank angle	deg.
θ_i	Start of combustion	deg.
θ_d	Combustion duration	deg.
k_g	thermal conductivity of gas	W/m-K
Nu	Nusselt number	–
$h_c(\theta)$	Convective heat transfer coefficient	W/m²-K
$h_r(\theta)$	Radiative heat transfer coefficient	W/m²-K

REFERENCES

1. J. B. Heywood, *Internal Combustion Engine Fundamentals*, Second Edition, 2018, Accessed: Apr. 20, 2021. [Online]. Available: https://www.accessengineeringlibrary.com/content/book/9781260116106.
2. G. P. Blair, "Design and simulation of four-stroke engines." Aug. 15, 1999, Accessed: Apr. 20, 2021. [Online]. Available: https://www.sae.org/publications/technical-papers/content/R-186/.
3. V. Ganesan, *Computer Simulation of Compression-Ignition Engine Processes*. Universities Press, 2000.
4. M. N. Nabi, M. Rasul, and P. Gudimetla, "Modelling and simulation of performance and combustion characteristics of diesel engine," *Energy Procedia*, vol. 160, no. December 2018, pp. 662–669, 2019, doi: 10.1016/j.egypro.2019.02.219.
5. T. K. Gogoi and D. C. Baruah, "A cycle simulation model for predicting the performance of a diesel engine fuelled by diesel and biodiesel blends," *Energy*, vol. 35, no. 3, pp. 1317–1323, Mar. 2010, doi: 10.1016/j.energy.2009.11.014.
6. S. Awad, E. G. Varuvel, K. Loubar, and M. Tazerout, "Single zone combustion modeling of biodiesel from wastes in diesel engine," *Fuel*, vol. 106, pp. 558–568, 2013, doi: 10.1016/j.fuel.2012.11.051.

7. V. Hariram and R. Bharathwaaj, "Application of zero-dimensional thermodynamic model for predicting combustion parameters of CI engine fuelled with biodiesel-diesel blends," *Alexandria Eng. J.*, vol. 55, no. 4, pp. 3345–3354, Dec. 2016, doi: 10.1016/j.aej.2016.08.021.

8. C. R. Stone and D. I. Green-Armytage, "Comparison of methods for the calculation of mass fraction burnt from engine pressure-time diagrams.," *Proc. Inst. Mech. Eng. Part D, Transp. Eng.*, vol. 201, no. D1, pp. 61–67, Feb. 1987, doi: 10.1243/pime_proc_1987_201_158_02.

9. R. K. Maurya, *Reciprocating Engine Combustion Diagnostics*. Cham: Springer International Publishing, 2019.

10. P. S. Gautam, P. K. Vishnoi, and V. K. Gupta, "A single zone thermodynamic simulation model for predicting the combustion and performance characteristics of a CI engine and its validation using statistical analysis," *Fuel*, vol. 315, p. 123285, May 2022, doi: 10.1016/j.fuel.2022.123285.

11. P. S. Gautam, P. K. Vishnoi, P. Maheshwari, T. S. Samant, and V. K. Gupta, "Experimental analysis and theoretical validation of C.I. engine performance and combustion parameters using zero-dimensional mathematical model fuelled with biodiesel and diesel blends," *{IOP} Conf. Ser. Mater. Sci. Eng.*, vol. 1168, no. 1, p. 12018, Jul. 2021, doi: 10.1088/1757-899x/1168/1/012018.

12. M. N. Nabi and M. G. Rasul, "One-dimensional thermodynamic model development for engine performance, combustion and emissions analysis using diesel and two paraffin fuels," *Energy Procedia*, vol. 156, no. September 2018, pp. 259–265, 2019, doi: 10.1016/j.egypro.2018.11.139.

13. P. K. Vishnoi, P. S. Gautam, P. Maheshwari, T. S. Samant, and V. K. Gupta, "Experimental evaluation of diesel engine operating with A {TERNARY} {BLEND} ({BIODIESEL}-{DIESEL}-{ETHANOL})," *{IOP} Conf. Ser. Mater. Sci. Eng.*, vol. 1168, no. 1, p. 12017, Jul. 2021, doi: 10.1088/1757-899x/1168/1/012017.

14. P. S. Gautam, P. K. Vishnoi, and V. K. Gupta, "The effect of water emulsified diesel on combustion, performance and emission characteristics of diesel engine," *Mater. Today Proc.*, no. xxxx, pp. 1–7, 2021, doi: 10.1016/j.matpr.2021.10.485.

15. P. K. Vishnoi, P. S. Gautam, and V. K. Gupta, "Impact of using n-pentanol as a co-solvent with diesel-methanol blends on combustion, performance and emissions of CI engine," *Int. J. Ambient Energy*, pp. 1–27, Dec. 2021, doi: 10.1080/01430750.2021.2013940.

Chapter 6

Analysis of hydrodynamic efficiency on a rectangular based OWC

A numerical simulation approach

Sanjeev Ranjan, Akshay Nitin Dorle and Pradip Deb Roy

National Institute of Technology, Silchar, India

CONTENTS

6.1 INTRODUCTION

In the field of wave energy conversion, the OWC technology is one of the most well-known devices in use. It comprises a structure that is partly immersed in water and has an air pocket trapped inside it that is above the free surface of the water. The water column inside the chamber oscillates as a result of the waves impinging on it. In response to the oscillating movement of the free surface inside of the pneumatic chamber, a piston is created by the water column within. Due to the velocity of the incident wave, the air within the chamber is compressed.

Based on the above fact, multiple researchers carried out various studies at different times, like floating OWC and fixed OWC. Wang et al. [1] validated numerical and experimental results. Delauré and Lewis [2] reported that the 3-D numerical model OWC device simulates the first-order boundary element methods (BEM). The authors appropriately designed the OWC and discuss. Count and Evans [3] work on NWT by coupling the 3-D boundary integral approach. Iturrioz et al. [4] applied the VOF approach to a fixed 3-D model and a floating OWC model, and the results were confirmed against the experimental data. According to Nunes et al. [5], a numerical analysis of an offshore OWC device focuses on increasing the device's energy extraction efficiency. Rezanejad et al. [6] compared dual-chamber OWC and single-chamber systems at a stepped bottom condition. Gkikas and Athanassoulis [7] used the identification approach used to work on a nonlinear system. Ning et al. [8] analyzed 2-D NWT by BEM. Falcão et al. [9] developed the hydrodynamic efficiency of an OWC.

Teixeira et al. [10] installed the OWC device on the shoreline using the Fluinco numerical model to simulate. Marjani et al. [11] developed an OWC system to anticipate the flow characteristics in the chamber. Iturrioz et al. [12] use stationary OWC for the simulation. Zhang et al. [13] studied the wave profile and gauge pressure inside the OWC. Luo et al. [14] studied the efficiency of OWC by PTO system. López et al. [15] used the RANS-VOF method to determine the efficiency of an OWC. Çelik et al. [16] determined the average surface fluctuation inside the chamber.

This research explores the effectiveness of a rectangular-based OWC based on the orifice and relative opening ratio in terms of orifice efficiency. The model has been investigated using the ANSYS FLUENT program. The NSE is solved using the finite volume approach. The VOF approach is used to locate and track the free surface flows. The numerical wave tank (NWT) in three dimensions (3-D) has been built to determine hydrodynamic efficiency. The findings are corroborated by the data from the experiments. The primary benefit of doing numerical analysis is the visualization of the flow around and inside the OWC, which gives a better understanding of the phenomena and may aid in the improvement of the design of the OWC for increased efficiency and effectiveness. It is possible to get a considerably more thorough picture of OWC performance when the cumulative impact of adjusting the orifice diameter and the front opening is taken into account.

6.2 NUMERICAL MODELING

6.2.1 Numerical model

The three-dimensional (3-D) NWT is rectangular with various boundary conditions. The boundary conditions implemented are shown in Figure 6.1. The inflow mechanism is used to generate waves at the inlet border, and it is quite effective. The top boundary is defined as a pressure outlet that is considered open to the atmosphere. The tank's bottom boundary and right boundary are set as no-slip conditions due to a solid wall. The velocity of fluid near the solid wall is zero due to the no-slip boundary condition. Cartesian coordinates are used in the x-z plane. The z-axis is directed vertically upward from the still water level, whereas the x-axis is defined

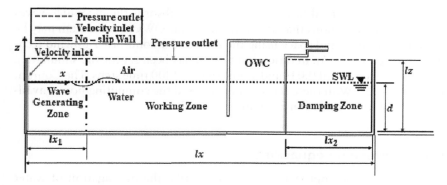

Figure 6.1 Definition sketch showing various zones and boundary conditions.

positively in the direction of wave propagation (SWL). The y-direction is taken perpendicular to the x-z plane pointing inwards to the plane of the paper.

The computational domain has been divided into three sections: a) wave generating zone (l_{x1} = 2m), b) working zone (l_{x2} = 12m), and c) damping zone (l_{x3} = 4m). The overall size of the NWT is l_x = 18m and l_z = 0.85m. In this study, the inflow method generates second-order stokes wave of time period T = 1.8 sec and wavelength L = 3.8548m. The structure grid has been utilized to discretize the computational domain. The mesh refinement has been implemented to capture the free surface in the fluid domain accurately. Figure 6.3 represents the uniformly structured grid mesh. The present study performed the grid convergence test to check the grid system under different grid sizes. Table 6.1 shows the four different meshes. Mesh sizes decrease gradually from Mesh 1 to Mesh 4. The nodes are 109467, 221076, 330267, and 420031 used for grid tests. It has been observed from the test results that no difference was found between Mesh 3 and Mesh 4 by reducing the size of the meshes from Mesh 3 to Mesh 4. As a result, the Mesh 3 system provides excellent performance and should be considered for future simulation.

This mesh size is selected because Gaspar et al. [17] and Luo et al. [14] have evaluated various sizes of mesh and showed that this size is appropriate for getting accurate results. The mesh gradually changes from fine to

Table 6.1 Mesh size parameters

Mesh	Δx (m)	Δy (m)	Δz (m)	Nodes
1	L/50	0.0625	H/10	109467
2	L/60	0.0500	H/15	221076
3	L/70	0.0275	H/20	330267
4	L/80	0.0125	H/25	420031

coarse in the numerical beach zone. This zone absorbs the incoming waves and prohibits reflections that affect the flow near OWC. Momentum may be discretized in time and space using the first- and second-order upwind techniques, respectively. An iteration limit of 30 has been set for the simulation, with a time step size of 0.01s and a time step of 2000 time steps. The Courant number is 0.25. With these settings in place and the computing power available, it took the simulation 20 hours to complete.

6.2.2 Governing equations

The finite volume approach is used to describe the propagation of water waves in a NWT, which is available as commercial software called ANSYS Fluent (FVM). In this model, irrational flow behavior is considered, along with the absence of surface tension and an ambient gauge pressure of zero. The NSE and the mass continuity equations (MCE) are employed as the governing equations of an unsteady, incompressible flow in the current situation. In Equation (6.2), the viscous term has been dropped since the inviscid flow model was used in the numerical solution.

$$\nabla(\vec{V}) = 0 \tag{6.1}$$

$$\frac{\partial}{\partial t}(\rho\vec{V}) + \nabla(\rho\vec{V}\vec{V}) = -\nabla P + \rho\vec{g} \tag{6.2}$$

Time (t) is measured in seconds, whereas pressure (p) is measured in Pascal, velocity vector $(\vec{V})$ in meter per second and density (ρ) is measured in kilograms per cubic meter.

Two or more non-interpenetrating fluids (or phases) are modeled in the VOF method as described by Hirt and Nichols [18]. A cell's volume fraction function α_q is defined as the percentage of its entire volume occupied by the qth phase. Here, the range of α_q is $0 \leq \alpha_q \leq 1$. The following equations are employed to determine α_q:

$$\frac{\partial\alpha_q}{\partial t} + \nabla(\alpha_q) = 0 \tag{6.3}$$

$$\sum_{q=1}^{2}\alpha_q = 1 \tag{6.4}$$

The two-phase mixture density is calculated based on the volume fraction as follows:

$$\rho = \alpha_q\rho_w + (1-\alpha_q)\rho_a \tag{6.5}$$

where $\rho_a = 1.225$ kg/m^3 and $\rho_w = 998.2$ kg/m^3 are the air and water density, respectively.

6.2.3 Hydrodynamic efficiency

Hydrodynamic efficiency (η) is defined as output power (P_{out}) to input power (P_{in}). Bouali and Larbi [19].

$$\eta = \frac{P_{out}}{P_{in}} \tag{6.6}$$

Equation (6.6) [20] gives the output power and Equation (6.7) [21] for the second-order stokes wave per unit width gives the input power:

$$P_{out} = \frac{1}{T} \int_0^T p(t)\, S_w v(t)\, dt \tag{6.7}$$

$$P_{in} = \frac{1}{8} \rho g H^2 \frac{\omega}{k} \left(\frac{1}{2} \left(1 + \frac{2kd}{\sinh(2kd)} \right) \right) \left(1 + \frac{9}{64} \frac{H^2}{k^4 d^6} \right) \tag{6.8}$$

where $p(t)$ = inside air chamber pressure, $v(t)$ = free surface velocity inside the chamber, S_w = sectional area of OWC, H (wave height), d (depth of water), k (wave number), wave angular frequency (ω), and T (wave period).

6.2.4 Geometry

Figure 6.2 shows the geometry of the OWC has been modeled according to the model presented experimentally in Celik et al. [22]. Ansys Design Modeler has been used to create a 3D CAD model. The geometry comprises

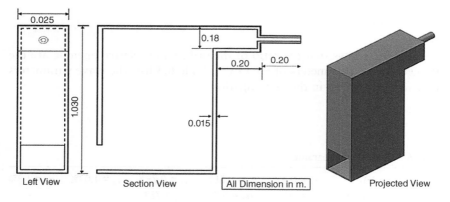

Figure 6.2 The geometry of the OWC.

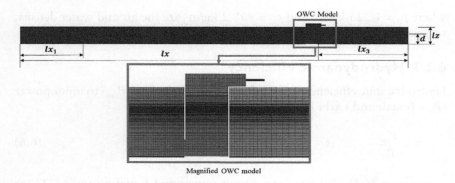

Figure 6.3 Computational mesh in the numerical domain.

the computational domain, which includes the wave tank and the OWC. Figure 6.2 shows the 3-D OWC model's dimensions. This model has been incorporated in the NWT, which has a length of $l_x = 18$ m, a width of $l_y = 0.26$ m, and a height of $l_z = 0.85$ m. The intake of the tank is 13.25 m from the OWC observation station. After the OWC, a zone having a length of 4 m is created to provide sufficient damping. This ensures no reflections from the back wall since reflections can cause unnecessary errors in calculations.

The definition of two geometrical parameters used in this study is as follows:

Orifice ratio (τ): It's the proportion of the orifice's cross-sectional area (S_o) to the free surface area (S_w) within the OWC.

$$\tau = S_o / S_w$$

Relative opening (α): It is the ratio of the height of opening (x) in the front wall of OWC to the depth of water (d) in the tank.

$$\alpha = x/d$$

Table 6.2 shows that the ratio of the orifice and relative opening are the two geometric parameters of the OWC. Table 6.3 lists the wave parameters that were employed in this investigation.

Table 6.2 Geometrical parameters

Orifice ratio (τ)	$\tau_1 = 0.4\%$	$\tau_2 = 0.58\%$	$\tau_3 = 0.79\%$	$\tau_4 = 1.03\%$	$\tau_5 = 1.3\%$
Relative opening (α)	$\alpha_1 = 33\%$	$\alpha_2 = 50\%$	$\alpha_3 = 75\%$	–	–

Table 6.3 Geometrical parameters

H	L	T	H/L	d/L
0.07 m	3.8548 m	1.8 s	0.02	0.16

6.3 MODEL VALIDATION

In order to validate the current numerical model, it was compared to the experimental model provided by Celik et al. [22]. The numerical data are assumed the same as the experimental data. Five orifice ratios (τ) are considered for the validation with a single relative openings $\alpha_3 = 75\%$ and wave steepness $H/L = 0.02$. Table 6.3 lists the wave parameters used in this study. Figure 6.4 shows the hydrodynamic efficiency (η) versus orifice ratio (τ) for a constant relative opening $\alpha_3 = 75\%$. The numerical outputs excellently correlate with experimental data, as can be shown in the figure. The maximum and minimum errors in the efficiency of the numerical model and experimental model are 4.2% and 0.07%, respectively. Hence, the present CFD model is appropriate for further study.

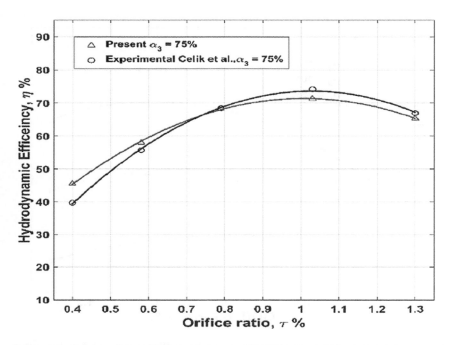

Figure 6.4 Comparing of hydrodynamic efficiency between the present and experimental models at $\alpha_3 = 75\%$ and $H/L = 0.02$.

6.4 RESULTS AND DISCUSSION

6.4.1 Effect of orifice ratio

Turbine damping is seen in both extreme scenarios. The first is zero orifice diameters, i.e., very high damping, and the second is maximum orifice diameter, i.e., zero damping. Here, pressure and velocity terms are zero according to Equation (6.7). Hence, pneumatic power will also be zero. Hence, the size of the orifice is varied to get the optimum absorbing power from the incident waves of the system under consideration. A second-degree polynomial curve fitting has been used in this study to better represent the performance characteristics, with the correlation coefficient $R^2 = 0.9251$.

Figure 6.5 describes hydrodynamic efficiency (η) versus orifice ratio (τ) at $H/L = 0.02$ under various values of relative openings (α). Figure 6.5 shows that the shape of the graph follows a concave behavior. It has been observed from Figure 6.5 that negligible efficiency is found at a low orifice ratio. At the smallest orifice diameter, generated air pressure within the OWC is very high due to high damping effect, obstructing the free surface movement within the OWC. Hence, less power is extracted. With the increases in orifice ratio, the hydrodynamic efficiency reaches a maxima

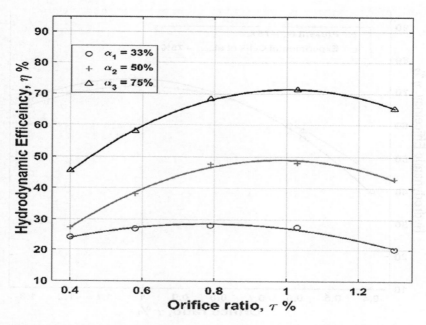

Figure 6.5 Hydrodynamic efficiency (η) versus orifice ratio (τ) for the wave steepness $H/L = 0.02$ at $d = 0.6$ m.

value and decreases for further increases in the orifice ratio. Hydrodynamic efficiency becomes a maximum value at $\tau = 1.03\%$ for all values of relative opening. This ratio of the orifice is the optimal orifice ratio (damping value) for $H/L = 0.02$.

Figure 6.6 illustrates the further analysis of mean air pressure and mean fluctuating velocity within the OWC at a) $\tau_1 = 0.40\%$, b) $\tau_2 = 0.58\%$, c) $\tau_3 = 0.79\%$, and d) $\tau_4 = 1.03\%$ for a single $\alpha_3 = 75\%$ and $H/L = 0.02$. We know from Equation (6.7) that the efficiency depends mainly on two factors: (1) air pressure in the chamber and (2) vertical velocity of the free surface in the chamber.

Figure 6.6(a) reveals that air pressure is high due to the low orifice ratio, which tightened the free surface velocity. Hence, efficiency here becomes low (Figure 6.5). Figure 6.6(d) shows that pressure and vertical velocity are optimal at a high orifice ratio. Hence, optimization of damping generates for all values of relative openings at orifice ratio $\tau_4 = 1.03\%$ and $H/L = 0.02$.

6.4.2 Effect of relative opening

With the OWC, the relative openness is critical to the overall operation of the device. The present study investigates hydrodynamic efficiency at three relative openings (α) and five different orifice ratios (τ). Figure 6.7 depicts the plots of hydrodynamic efficiency vs relative openness for five various orifice ratios (τ) and for the $H/L = 0.02$ condition, respectively.

As seen in Figure 6.7, the efficiency of the OWC rises practically linearly as the relative openness of the OWC increases. The minimum efficiency is found at the smallest relative opening of α_1 and maximum efficiency at α_3, which is the largest opening in this study. As we know, the kinetic energy of the propagation wave on the free surface is maximum and minimum at the bottom. So for the smallest relative opening α_1 maximum energy will reflect due to the front wall, which cannot be utilized. As the opening is increased, more wave energy enters the chamber, which increases the hydrodynamic efficiency.

6.5 CONCLUSION

The influence of the orifice ratio and relative opening on the hydrodynamic efficiency of a rectangular-based OWC was explored in the current numerical model. In order to create the appropriate nonlinear waves in a viscous NWT, the inflow velocity technique and the Dirichlet boundary condition are utilized. To further understand the influence of orifice ratio and relative opening, five orifice ratios (τ) and three relative openings (α) are investigated for $d = 0.6m$ and $H/L = 0.02$ for the following parameters: A comparison is made between the numerical model and the experimental model.

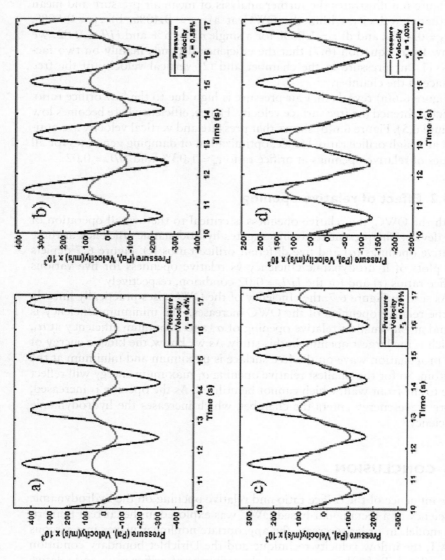

Figure 6.6 Pressure and water column velocity versus time at a) $\tau_1 = 0.40\%$, b) $\tau_2 = 0.58\%$, c) $\tau_3 = 0.79\%$, and d) $\tau_4 = 1.03\%$ for a single $\alpha_3 = 75\%$ and $H/L = 0.02$.

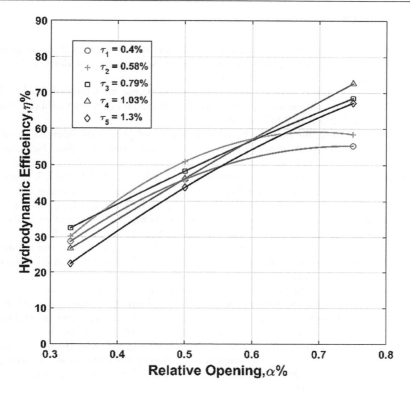

Figure 6.7 Hydrodynamic efficiency (η) versus relative opening (α) at $H/L = 0.02$ and at $d = 0.6$ m.

Itdemonstrates that the numerical data and the experimental data are very well matched. The current model accepted the results that follow.

- The hydrodynamic efficiency of the system follows a concave path for each relative opening as the orifice ratio rises. $\tau_4 = 1.03\%$ for $\alpha_3 = 75\%$ is the system's maximum efficiency.
- Chamber pressure increases with the lower value of orifice ratio (τ).
- With the increase in the relative opening, the hydrodynamic efficiency increases linearly, with the maximum efficiency being at $\alpha_3 = 75\%$.
- As the relative opening increases, more kinetic energy is utilized from the incoming wave by the OWC, and therefore the hydrodynamic efficiency increases with the relative opening.
- For a particular design of the OWC chamber, it is extremely important to determine the optimal value of turbine damping to maximize the energy extraction from the wave.
- The relative opening should be as large as possible as long as air leakage into the chamber is avoided for maximum power extraction.

REFERENCES

1. D. J. Wang, M. Katory, and Y. S. Li, Analytical and experimental investigation on the hydrodynamic performance of onshore wave-power devices. *Ocean Engineering*, 29, 2002, pp. 871–885.
2. Y. M. C. Delauré, and A. Lewis, 3-D hydrodynamic modelling of fixed oscillating water column wave power plant by a boundary element method. *Ocean Engineering*, 30(3), 2003, pp. 309–330.
3. B. M. Count, and D. V. Evans, The influence of projecting sidewalls on the hydrodynamic performance of wave-energy devices. *Journal of fluid Mechanics*, 145, 1984, pp. 361–376.
4. A. Iturrioz, R. Guanche, J. A. Armesto, M. A. Alves, C. Vidal, and I. J. Losada, Time-domain modeling of a fixed detached oscillating water column towards a floating multi-chamber device. *Ocean Engineering*, 76, 2014, pp. 65–74.
5. G. Nunes, D. Valério, P. Beirão, and J. S. Da Costa, Modelling and control of a wave energy converter. *Renewable Energy*, 36, 2011, pp. 1913–1921.
6. K. Rezanejad, J. Bhattacharjee, and C. G. Soares, Analytical and numerical study of dual-chamber oscillating water columns on stepped bottom. *Renewable Energy*, 75, 2015, pp. 272–282.
7. G. D. Gkikas, and G. A. Athanassoulis, Development of a novel nonlinear system identification scheme for the pressure fluctuation inside an oscillating water column-wave energy converter part I: theoretical background and harmonic excitation case. *Ocean Engineering*, 80, 2014, pp. 84–99.
8. D. Z. Ning, J. Shi, Q. P. Zou, and B. Teng, Investigation of hydrodynamic performance of an OWC (oscillating water column) wave energy device using a fully nonlinear HOBEM (higher-order boundary element method). *Energy*, 83, 2015, pp. 177–188.
9. A. F. O. Falcão, J. C. C. Henriques, and J. J. Cândido, Dynamics and optimization of the OWC spar buoy wave energy converter. *Renewable Energy*, 48, 2012, pp. 369–381.
10. P. R. F. Teixeira, D. P. Davyt, E. Didier, and R. Ramalhais, Numerical simulation of an oscillating water column device using a code based on Navier-Stokes equations. *Energy*, 61, 2013, pp. 513–530.
11. A. E. Marjani, F. C. Ruiz, M. A. Rodriguez, and M. T. P. Santos, Numerical modelling in wave energy conversion systems. *Energy*, 33, 2008, pp. 1246–1253.
12. A. Iturrioz, R. Guanche, J. L. Lara, C. Vidal, and I. J. Losada, Validation of Open FOAM_ for oscillating water column three-dimensional modeling. *Ocean Engineering*, 107, 2015, pp. 222–236.
13. Y. Zhang, Q. P. Zou, and D. Greaves, Air-water two-phase flow modeling of hydrodynamic performance of an oscillating water column device. *Renewable Energy*, 41, 2012, pp. 159–170.
14. Y. Luo, J. R. Nader, P. Cooper, and S. P. Zhu, Nonlinear 2D analysis of the efficiency of fixed oscillating water column wave energy converter. *Renewable Energy*, 64, 2014, pp. 255–265.
15. I. López, B. Pereiras, F. Castro, and G. Iglesias, Optimization of turbine-induced damping for an OWC wave energy converter using a RANS-VOF numerical model. *Applied Energy*, 127, 2014, pp. 105–114.

16. A. Çelik, and A. Altunkaynak. Determination of damping coefficient experimentally and mathematical vibration modelling of OWC surface fluctuations. *Renewable Energy*, 147, 2020, pp. 1909–1920.
17. L. A. Gaspar, P. R. Teixeira, and E. Didier, Numerical analysis of the performance of two onshore oscillating water column wave energy converters at different chamber wall slopes. *Ocean Engineering*, 201, 2020, pp. 107–119.
18. W. C. Hirt, and D. B. Nichols, Volume of fluid (VOF) method for the dynamics of free boundaries. *Journal of Computational Physics*, 39(1), 1981, pp. 201–225.
19. B. Bouali, and S. Larbi, Sequential optimization and performance prediction of an oscillating water column wave energy converter. *Ocean Engineering*, 131, 2017, pp. 162–173.
20. M. T. Morris-Thomas, R. J. Irvin, and K. P. Thiagarajan, An investigation into the hydrodynamic efficiency of an oscillating water column, 2007, pp. 273–278.
21. B. Bouali, and S. Larbi, Contribution to the geometry optimization of an oscillating water column wave energy converter. *Energy Procedia*, 36, 2013, pp. 565–573.
22. A. Çelik, and A. Altunkaynak, Experimental investigations on the performance of a fixed-oscillating water column type wave energy converter. *Energy*, 188, 2019, pp. 116071.

Chapter 7

Segregation in size bi-dispersed system of active particles

Siddhant Mohapatra, Sahithya Pandula and Pallab Sinha Mahapatra

Indian Institute of Technology Madras, Chennai, India

CONTENTS

7.1 INTRODUCTION TO ACTIVE MATTER

Active matter has been a fascinating subject of research owing to its ubiquity in nature. The rich array of behavioral dynamics exhibited by these self-propelling systems holds immense significance in fields ranging from biophysics [1–2] to crowd dynamics [3] to swarm robotics [4]. Structural features such as cluster formation, motility-induced phase separation, and giant number fluctuations [5–6] make it imperative to understand the underlying principles governing this coordinated motion. Although there are individual differences among units of natural and synthesized active systems, most studies have focused on mono-dispersed systems of active entities. Many novel phenomena such as aggregation, crystallization, flocking, sedimentation, etc., have been discerned for such systems. Therefore, there is a paramount need to consider the diversity on the individual level, which has been tackled to some extent in the current work by incorporating bidispersity in the system. The exact dynamics of the information exchange in collective motion are not yet deciphered. However, numerous mathematical models pertaining to collective behavior have been developed over the years to emulate it. They can be categorized as (1) particle-based models, where the behavior of dry active matter is analyzed in the absence of hydrodynamic interactions, and the conservation of momentum and energy doesn't hold, and (2) continuum models for active suspensions or wet systems. Vicsek model [7], the simplest particle-based model, has been central to most active matter studies. In this model, the particles move with constant velocity in a medium without any hydrodynamic interactions. They interact and align themselves in the average direction of their neighbors' velocities. At high particle density and low noise strength, the system undergoes a transition from a disordered state to an ordered state where the particles move collectively in the same direction. This simple model was adapted for crowd dynamics by Helbing et al. [3] and for biological systems by Couzin et al. [8]. There have been many proposed extensions of the Vicsek model by inclusion of separation drive [9], extrinsic noise [10], averaging acceleration alongside velocity [11], to name a few. Different methods of neighborhood selection have also been proposed, such as the metric distance method of the Vicsek model, topological method [12], and Voronoi neighbor method [13] (a subset of the topological method). A decision model was superposed on the Vicsek model by Chaté et al. [14] where each particle has the option to either follow the standard alignment rule and move with the flock or move exactly opposite to the average heading of the flock. Another agent-based model is the Boids' flocking model [15], which considers all the entities as Boids' having a zoning structure and corresponding drives for each zone. Strömbom [16] defined a model considering only attraction drive and reported the occurrence of the mill (circular motion) as well as polarized motion (highly parallel motion). Grossman et al. [17] reported phase transitions in a self-propelled particle system using only inelastic inter-particle collision rule. Attractive and repulsion interactions among the particles have also been developed involving Lennard-Jones potential [18] and Quasi-Morse potential [19]. In a unique perspective to attraction and repulsion, Romanczuk et al. [20] described an escape-pursuit model where the agents are focused on the agents in front of them, while also moving away from the agents pursuing them. Szabo et al. [21] proposed a basic adhesion-repulsion model for the movement of cells in monolayers. Cucker and Smale [22] gave a mathematician's perspective to flocking and derived equations for the emergence of flocking in a group of organisms. They have also defined the critical limit for maintaining the integrity of the flock. However, the assumption of a noise-free environment and the absence of a separation parameter were certain shortcomings of the original model. These pitfalls were mended in subsequent works by Cucker and Mordecki [23] for flocking under noisy conditions and Cucker and Dong [24] for avoidance behavior in flocking. Apart from the above models, which mostly do not account for the hydrodynamic effects associated with the motion, continuum models such as Toner and Tu model [25] have also been developed. The details of the aforementioned models have been discussed in exhaustive detail in [26–29]. As of late, evolutionary models [30] and models based in network theory [31–32] have also come up and new-age concepts such as reinforcement learning have been put into use [33].

In this work, a force-based interpretation of the Vicsek model is considered to study the collective motion of size-differentiated active matter. Phase transitions, being quintessential to active matter, the continuous/discontinuous nature and the order of such phenomena have been a highly debated topic [7, 10, 34]. Differences in size, shape, mass, and motility affect the properties of the collective in binary mixtures. It has been reported that a mixture of size bi-dispersed active disks segregates in the absence of any adhesive interaction and can prevent the occurrence of crystallization [35]. Interaction among particles in such systems is highly dependent on the size ratio and volume fractions, leading to either homogeneity or motility-induced phase separation [36]. Controlling different phases formed in a binary mixture of active and passive particles has been found to increase the efficacy of self-powered drug delivery systems, cleaning polluted water, and many other microfluidic processes. Segregation, the primary proponent in these cases, is generally driven by differences in parameters such as density, composition, motility, etc. Mones et al. [37] showed the segregation of active particles induced by differential adhesion, while Yang et al. [38] reported the segregation in case of pure repulsion and Costanzo et al. [39] construed the motility-induced segregation of particles inside a microchannel. Agrawal and Mahapatra [40] studied the relation between segregation in a mixture of motile and passive particles and coordination, concluding that a mono-disperse system is preferable to bidispersity as bidispersity promotes a mixed state of the system. This work aims to investigate the complex behavior of a system constituting active bi-dispersed particles in a confined domain by analyzing the phases observed and quantifying the mixing/segregation of the particles.

7.2 NUMERICAL METHODOLOGY

A discrete element model is defined as consisting of multiple particles, which are tracked in a Langrangian framework. The model is based on three types of forces: the inter-particle force $\vec{F}_{pp}$, the self-propelled force $\vec{F}_{sp}$, and the coordination force $\vec{F}_c$. The total force acting on the ith particle is given by Equation 7.1.

$$m_i \frac{d\vec{v}_{p,i}}{dt} = \vec{F}_{sp} + \vec{F}_{pp} + \vec{F}_c \qquad (7.1)$$

where m_i and $\vec{v}_{p,i}$ are the mass and the velocity of the ith particle, respectively.

The inter-particle force $\vec{F}_{pp}$ acts in a pair-wise manner on the particles when two particles overlap with each other to separate the two particles. Equation 7.2, showing the formulation for this force, is derived from Hertzian contact theory for two cylinders with parallel axes [41].

$$\vec{F}_{pp} = \begin{cases} k_n \vec{\delta}, |\vec{\delta}| < 0 \\ \vec{0}, |\vec{\delta}| \geq 0 \end{cases} \qquad (7.2)$$

where $\vec{\delta} = \left(|\vec{r}_i - \vec{r}_j| - \dfrac{d_i + d_j}{2} \right) \hat{r}_{ij}$ is the separation between the two particles i

and j and $\hat{r}_{ij}$ is the direction away from the overlap.

The coordination force [42] is a reiteration of the Vicsek alignment rule where the particle endeavors to align itself to its neighbors. The formulation (see Equation 7.3) has been derived from the drag on a sphere in a Stokes flow regime. A delta-correlated Gaussian random noise term with mean zero and variance $\sigma^2/12$ is also introduced into this force to include the effects of imperfect alignment. The parameter σ is fixed at 1 for all the simulation as this work is not aimed toward studying the effect of noise on segregation dynamics. The addition of noise has been carried out in a similar fashion as that of Agrawal and Mahapatra [43].

$$\vec{F}_c = C_v d_i (\vec{v}_i - \vec{v}_{p,i}) \tag{7.3}$$

where $\vec{v}_i$ is the velocity of the fluid surrounding particle i calculated from the weighted average of the neighboring particle velocities due to the sparsity of fluid content.

The self-propulsion force [18] (see Equation 7.4) is by virtue of the activity/motility of the particle. It causes propulsion of the particles by lending energy from an assumed infinite energy source, and hence is the major contributor to the departure of active matter systems from equilibrium.

$$\vec{F}_{sp} = m_i (\beta_i - \alpha \, |\vec{v}_{p,i}\,|^2) \hat{v}_{p,i} \tag{7.4}$$

where β_i is the thrust coefficient for particle i and α is the Rayleigh friction factor to prevent unbounded acceleration.

Particles of both sizes are placed following a uniform random distribution within a square enclosure with reflective boundaries as shown in Figure. 7.1 and the net force on each particle is calculated from Equation 7.1, which is then numerically integrated using the Velocity-Verlet scheme to obtain the velocity and the displacement for each time step. Due to its lower time complexity, a linked-list algorithm has been implemented to build the Verlet list (neighbor list). Table 7.1 lists all the parameters and their respective units used in the model.

7.3 RESULTS AND DISCUSSION

The system is analyzed in the purview of two non-dimensional values: non-dimensional time $\tau = t/\sqrt{L/\beta}$ and $\chi = \dfrac{C_v L \sqrt{L}}{(m_s + m_l)\beta}$ (m_s and m_l being the

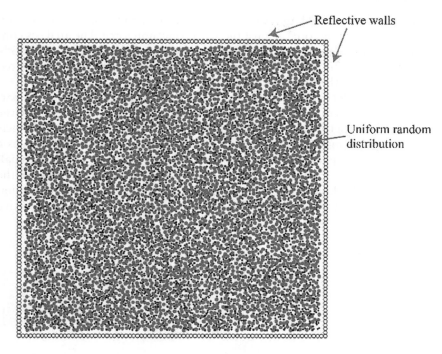

Figure 7.1 Representative image of the initial state of the system.

Table 7.1 Nomenclature of the symbols used in the work

Symbol	Meaning	Unit
α	Rayleigh friction factor	[m^{-1}]
β	Thrust coefficient	[ms^{-2}]
χ	Non-dimensional coordination coefficient	–
δ	Extent of overlap	[m]
$\langle r \rangle$	Rotational order parameter	–
η	Segregation index	–
ρ	Number fraction of small particles	–
τ	Non-dimensional time	–
ϱ	Number ratio of small particles to large particles	–
C_v	Coordination coefficient	[kg m^{-1} s^{-1}]
D	Diameter of particle	[m]
k_n	Elastic constant	[kg s^{-2}]
m_l	Mass of large particle	[kg]
m_s	Mass of small particle	[kg]
v_p	Velocity of particle	[ms^{-1}]

mass of the small and large particles, respectively) defined as the comparison of the coordination force against the self-propulsion force. The analysis considers two different packing fractions of the closed domain, i.e., $\Phi = 33\%$ and $\Phi = 60\%$. The ratio of the number of the small to large particles is fixed at $\varphi = 1.0$, and the ratio of radii for the same is fixed at $\varrho = 0.5$.

The parameter χ plays a vital role in the behavioral dynamics of the particles in the system. At low coordination (see Figure 7.2, $\chi = 73$), both large and small particles don't align with each other strongly. As a result, their movement has more individuality because of their $\vec{F}_{sp}$, which shows up as a diffused milling phase with no clear visual segregation. At $\chi = 725$, the small particles (colored black) are found to accumulate at the corners of the domain, while the large particles (colored blue) move in a milling motion around a square empty core. The mill is a frequently observed collective

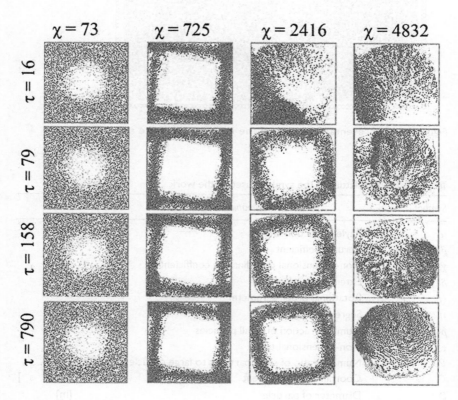

Figure 7.2 The snapshots of the system at different time instances $\tau = t / \sqrt{\dfrac{L}{\beta}}$ and non-dimensional coordination coefficient $\chi = \dfrac{C_v L \sqrt{L}}{(m_s + m_l)\beta}$ are shown for packing fraction $\Phi = 33\%$ and number ratio $\varphi = 1.0$.

phenomenon in a confined domain [5]. Further increase in coordination (χ) leads to the small particles moving in a similar mill as the large particles (see Figure 7.2, $\chi = 2416$). In the case of $\chi = 725$, the small particles didn't have enough motile force or coordination force to get themselves out of the corners. However, at $\chi = 2416$, the coordination force is sufficiently high so that the small particles align and follow the large particles and don't get entrapped in the domain corners. At very high coordination values $(\chi = 4832)$, the overall motion of the system transitions from milling to an oscillatory state. The previously empty core collapses and particles fill up that void. The reflective boundaries play a major role in this case as it is responsible for the change in the direction of the motion of the flock. On reaching a stable state $(\tau = 790)$, it can be observed that the small particles form a layer outside the large particles' core. It is also noted that with an increase in χ, the time required to reach a stable state increases.

The packing fraction Φ also has a major impact on the pattern formation in active matter systems, as evidenced by Yang et al. [38]. Figure 7.3 elucidates the contrast in the self-organization among particles at two different Φ. The first difference is related to the core size. The size of the core in the case of milling motion is significantly lesser at $\Phi = 60\%$ when compared to $\Phi = 33\%$. This is inevitable due to the increased space required for the increased number of particles and the inter-particle force that disallows the particles from overlapping. It can also be observed that in the case of $\chi = 725$, $\Phi = 33\%$ has a square-shaped core, while $\Phi = 60\%$ has a circular core whenever milling is found to occur. Similarly, for $\chi = 4832$, the particles form an undivided cluster at the center, which bounces between the walls at lower Φ, whereas, at higher Φ, they slowly oscillate between the walls taking a distorted square shape.

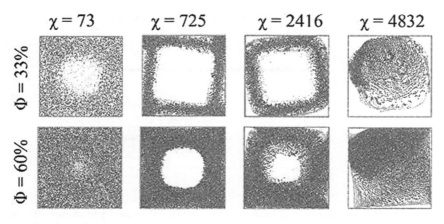

Figure 7.3 The stable state of the system is illustrated for different Φ and χ; $\varphi = 1.0$.

The motion of the particles observed at $\chi = 73$, $\chi = 725$, and $\chi = 2416$ are variants of the mill. To differentiate the motion among these cases, a rotational order parameter $\langle r \rangle = \dfrac{1}{N} \left| \displaystyle\sum_{i=1}^{N} \dfrac{\vec{r}_{i,cm} \times \vec{v}_{p,i}}{|\vec{r}_{i,cm} \times \vec{v}_{p,i}|} \right|$ has been defined a priori [43] by computing the normalized angular momentum of the system, where $\vec{r}_{i,cm} = \vec{r}_{p,i} - \vec{r}_{cm}$ is the position vector of the particle relative to the instantaneous center of mass of the system. The orientational order in the case of milling is represented through this order parameter such that the higher the value of $\langle r \rangle$, more orderly is the milling. Figure 7.4 delineates the temporal variation of the rotational order parameter for different cases of χ and Φ. With the increase in coordination χ, the rotational order parameter is found

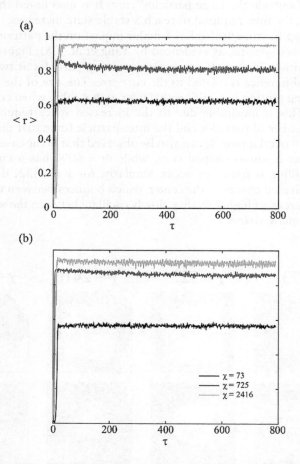

Figure 7.4 The temporal evolution of the rotational order parameter $\langle r \rangle$ is presented for different coordination levels χ at packing fractions (a) $\Phi = 33\%$ and (b) $\Phi = 60\%$; $\varphi = 1.0$.

to increase. While quite evident for a monodisperse system, this behavior is not intuitive in a bidispersed system. Despite the momentum difference between the differently sized particles (causing segregation, discussed later), which increases with χ, the angular velocity of the system shows an opposite trend. Similar behavior is observed in Figure 7.4(b) for higher packing fraction, although the values of the rotational order parameter are comparatively lower than at $\Phi = 33\%$.

The second difference is concerned with the mixing of the two kinds of particles in the domain. To discern this mixing more keenly, a segregation index $\eta = \sqrt{\sum_{i=1}^{N} \frac{(\rho_i - \langle \rho \rangle)^2}{N-1}}$ has been defined [40, 44], where the entire domain is divided into N small cells, ρ_i is the number fraction of small particles (i.e., ratio of a number of small particles to the total number of particles) in the ith cell and $\langle \cdot \rangle$ stands for the ensemble average. This division of the domain is illustrated in Figure 7.5. Different cell sizes are considered, and segregation index is calculated for each case. Finally, the mean of the segregation indices for the different cell sizes is reported as η.

Figure 7.6 illustrates the temporal change in segregation index η in cases of different χ and Φ. It can be observed straightaway that for any value of χ, η is higher in the case of $\Phi = 33\%$ when compared to $\Phi = 60\%$. This, in turn, implies that there is a higher extent of mixing observed at higher packing fractions. At $\chi = 73$, the segregation index remains almost unchanged throughout the simulation time, whereas at $\chi = 725$ and $\chi = 2416$, the segregation index settles down after a certain extent of time. The increase in the segregation index with the increase in alignment χ suggests that χ is a proponent of segregation. This anomaly can be explained by examining the

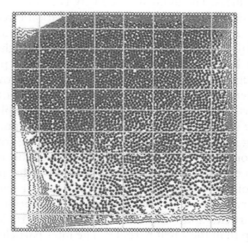

Figure 7.5 Domain division for calculating segregation index η.

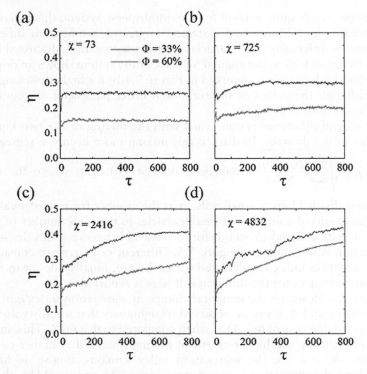

Figure 7.6 The temporal variation of segregation index η is displayed for different Φ and χ; $\varphi = 1.0$.

differences in the momenta of the two particle species, brought out by the difference in the mass. The momenta disparity is contributed jointly by the self-propulsion (dependent on mass) and coordination force (dependent on diameter) terms. As a result, even though there is coordination among all particles, eventually, the momenta difference is high enough that the smaller particles cannot keep up with the motion of the large particles and are pushed to the edges of the mill or cluster. It is also discerned, from Figure 7.5, that the difference between the segregation indices for the two packing fraction values is much more prominent in the milling phase than in the oscillatory phase ($\chi = 4832$). This implies that milling promotes segregation at lower packing fraction and mixing at higher packing fraction, while the oscillatory motion of the particles promotes segregation regardless of the packing fraction of the system. A possible reason for this peculiarity might be the presence of the inter-particle force. This force is responsible for causing the oscillatory behavior of the particles as the interior particles collide with the wall particles and bounce back. The large particles have a higher extent of overlap than the small particles. The inter-particle force is directly proportional to the extent of overlap and independent of mass, as mentioned in Equation 7.2. The differential application of this force on the two kinds of particles causes the smaller particles to be reflected with a lower thrust, causing the smaller particles to remain as an outer layer to the collective as can be seen in Figure 7.3. This phenomenon happens at any packing fraction for this phase.

7.4 CONCLUSION

In this study, a system of self-propelling particles with size-based bidispersity placed in a confined domain has been scrutinized on the coordination and packing fraction variation. The visual features of the complex behavior of the particles have been illustrated. The occurrence of two phases, the milling and oscillatory phases is observed on varying the degree of coordination. The differences in the milling behavior across cases have also been quantified with help of a rotational order parameter, which increases with coordination. In a few cases, it is observed that the size bidispersity causes the particles to segregate. To quantify the homogeneity/heterogeneity of the mixture, a segregation index is defined, and its temporal variation is analyzed under differing coordination and packing fraction conditions. It is found that there is an unchanging value of segregation at low levels of coordination and that milling behavior is found to cause an inverse relation between packing fraction and mixing. While increasing coordination does lead to segregation (due to momenta disparity), at the same time, the angular momenta of the particles are well aligned (i.e., the orientation of the particles are aligned), which makes for an interesting scenario. In the case of the oscillatory phase, it is seen that segregation occurs irrespective of the packing fraction. The segregation phenomenon can have major implications in cell sorting, where cells are differentiated based on their properties. However, the collective behavior in a bi-dispersed active matter system is a complex phenomenon and further work is necessary to understand it with more clarity. An in-depth analysis involving useful tools such as the Gini coefficient [38], mean square displacement [43], phase plots, and chaos analysis [45] will aid in having a firmer grasp of this subject matter.

REFERENCES

1. A. Be'er, B. Ilkanaiv, R. Gross, D. B. Kearns, S. Heidenreich, M. Bär, and G. Ariel, "A phase diagram for bacterial swarming." *Communication Physics*, vol. 3(1), pp. 1–8, 2020. https://doi.org/10.1038/s42005-020-0327-1

2. S. Mohapatra, and P. S. Mahapatra, "Confined system analysis of a predator-prey minimalistic model." *Scientific Reports*, vol. 9(1), pp. 11258, 2019. https://doi.org/10.1038/s41598-019-47603-9

3. D. Helbing, I. Farkas, and T. Vicsek, "Simulating dynamical features of escape panic." *Nature*, vol. 407(6803), pp. 487–490, 2000. https://doi.org/10.1038/35035023

4. Y. Yang, and M. A. Bevan, "Cargo capture and transport by colloidal swarms." *Science Advances*, vol. 6(4), pp. eaay7679, 2020. https://doi.org/10.1126/sciadv.aay7679

5. D. Armbruster, S. Motsch, and A. Thatcher, "Swarming in bounded domains." *Physica D: Nonlinear Phenomena*, vol. 344, pp. 58–67, 2017. https://doi.org/10.1016/j.physd.2016.11.009

6. G. Gonnella, D. Marenduzzo, A. Suma, and A. Tiribocchi, "Motility-induced phase separation and coarsening in active matter." *Comptes Rendus Physique*, vol. 16(3), pp. 316–331, 2015. https://doi.org/10.1016/j.crhy.2015.05.001

7. T. Vicsek, A. Czirók, E. Ben-Jacob, I. Cohen, and O. Shochet, "Novel type of phase transition in a system of self-driven particles." *Physical Review Letters*, vol. 75(6), pp. 1226, 1995. https://doi.org/10.1103/PhysRevLett.75.1226

8. I. D. Couzin, J. Krause, R. James, G. D. Ruxton, and N. R. Franks, "Collective memory and spatial sorting in animal groups." *Journal of Theoretical Biology*, vol. 218(1), pp. 1–11, 2002. https://doi.org/10.1006/jtbi.2002.3065

9. A. Czirók, E. Ben-Jacob, I. Cohen, and T. Vicsek, "Formation of complex bacterial colonies via self-generated vortices." *Physical Review E*, vol. 54(2), pp. 1791, 1996. https://doi.org/10.1103/PhysRevE.54.1791

10. G. Grégoire, and H. Chaté, "Onset of collective and cohesive motion." *Physical Review Letters*, vol. 92(2), pp. 025702, 2004. https://doi.org/10.1103/PhysRevLett.92.025702

11. P. Szabó, M. Nagy, and T. Vicsek, "Transitions in a self-propelled-particles model with coupling of accelerations." *Physical Review E*, vol. 79(2), pp. 021908, 2009. https://doi.org/10.1103/PhysRevE.79.021908

12. M. Ballerini, N. Cabibbo, R. Candelier, A. Cavagna, E. Cisbani, I. Giardina, V. Lecomte, A. Orlandi, G. Parisi, A. Procaccini, M. Viale, and V. Zdravkovic, "Interaction ruling animal collective behavior depends on topological rather than metric distance: Evidence from a field study." *Proceedings of the National Academy of Sciences*, vol. 105(4), pp. 1232–1237, 2008. https://doi.org/10.1073/pnas.0711437105

13. F. Ginelli, and H. Chaté, "Relevance of metric-free interactions in flocking phenomena." *Physical Review Letters*, vol. 105(16), pp. 168103, 2010. https://doi.org/10.1103/PhysRevLett.105.168103

14. H. Chaté, F. Ginelli, and R. Montagne, "Simple model for active nematics: Quasi-long-range order and giant fluctuations." *Physical Review Letters*, vol. 96(18), pp. 180602, 2006. https://doi.org/10.1103/PhysRevLett.96.180602

15. C. W. Reynolds, "Flocks, herds, and schools: A distributed behavioral model," in *Proceedings of the 14th Annual Conference on Computer Graphics and Interactive Techniques*, 1987, pp. 25–34. https://doi.org/10.1145/37402.37406

16. D. Strömbom, "Collective motion from local attraction." *Journal of Theoretical Biology*, vol. 283(1), pp. 145–151, 2011. https://doi.org/10.1016/j.jtbi.2011.05.019

17. D. Grossman, I. S. Aranson, and E. B. Jacob, "Emergence of agent swarm migration and vortex formation through inelastic collisions." *New Journal of Physics*, vol. 10(2), pp. 02303, 2008. https://doi.org/10.1088/1367-2630/10/2/023036

18. M. R. D'Orsogna, Y. L. Chuang, A. L. Bertozzi, and L. S. Chayes, "Self-propelled particles with soft-core interactions: Patterns, stability, and collapse." *Physical Review Letters*, vol. 96(10), pp. 104302, 2006. https://doi.org/10.1103/PhysRevLett.96.104302

19. J. A. Carrillo, S. Martin, and V. Panferov, "A new interaction potential for swarming models." *Physica D: Nonlinear Phenomena*, vol. 260, pp. 112–126, 2013. https://doi.org/10.1016/j.physd.2013.02.004

20. P. Romanczuk, I. D. Couzin, and L. Schimansky-Geier, "Collective motion due to individual escape and pursuit response." *Physical Review Letters*, vol. 102(1), pp. 010602, 2009. https://doi.org/10.1103/PhysRevLett.102.010602

21. B. Szabo, G. J. Szöllösi, B. Gönci, Z. Jurányi, D. Selmeczi, and T. Vicsek, "Phase transition in the collective migration of tissue cells: Experiment and model." *Physical Review E*, vol. 74(6), pp. 061908, 2006. https://doi.org/10.1103/PhysRevE.74.061908

22. F. Cucker, and S. Smale, "Emergent behavior in flocks." *IEEE Transactions on Automatic Control*, vol. 52(5), pp. 852–862, 2007. https://doi.org/10.1109/TAC.2007.895842

23. F. Cucker, and E. Mordecki, "Flocking in noisy environments." *Journal de mathématiques pures et appliquées*, vol. 89(3), pp. 278–296, 2008. https://doi.org/10.1016/j.matpur.2007.12.002

24. F. Cucker, and J. G. Dong, "Avoiding collisions in flocks." *IEEE Transactions on Automatic Control*, vol. 55(5), pp. 1238–1243, 2010. https://doi.org/10.1109/TAC.2010.2042355

25. J. Toner, and Y. Tu, "Flocks, herds, and schools: A quantitative theory of flocking." *Physical Review E*, vol. 58(4), pp. 4828, 1998. https://doi.org/10.1103/PhysRevE.58.4828

26. H. Chaté, "Dry aligning dilute active matter." *Annual Review of Condensed Matter Physics*, vol. 11, pp. 189–212, 2020. https://doi.org/10.1146/annurev-conmatphys-031119-050752

27. S. Ramaswamy, "The mechanics and statistics of active matter." *Annual Review of Condensed Matter Physics*, vol. 1(1), pp. 323–345, 2010. https://doi.org/10.1146/annurev-conmatphys-070909-104101

28. M. C. Marchetti, J. F. Joanny, S. Ramaswamy, T. B. Liverpool, J. Prost, M. Rao, and R. A. Simha, "Hydrodynamics of soft active matter." *Reviews of Modern Physics*, vol. 85(3), pp. 1143, 2013. https://doi.org/10.1103/RevModPhys.85.1143

29. T. Speck, "Collective forces in scalar active matter." *Soft Matter*, vol. 16(11), pp. 2652–2663, 2020. https://doi.org/10.1039/d0sm00176g

30. J. Demšar, C. K. Hemelrijk, H. Hildenbrandt, and I. L. Bajec, "Simulating predator attacks on schools: Evolving composite tactics." *Ecological Modelling* vol. 304, pp. 22–33, 2015. https://doi.org/10.1016/j.ecolmodel.2015.02.018

31. H. Shirado, F. W. Crawford, and N. A. Christakis, "Collective communication and behaviour in response to uncertain 'Danger' in network experiments." *Proceedings of the Royal Society A*, vol. 476(2237), pp. 20190685, 2020. https://doi.org/10.1098/rspa.2019.0685

32. E. Berekméri, I. Derényi, and A. Zafeiris, "Optimal structure of groups under exposure to fake news." *Applied Network Science*, vol. 4(1), pp. 1–13, 2019. https://doi.org/10.1007/s41109-019-0227-z

33. X. Wang, J. Cheng, and L. Wang, "A reinforcement learning-based predator-prey model." *Ecological Complexity*, vol. 42, pp. 100815, 2020. https://doi.org/10.1016/j.ecocom.2020.100815

34. M. Aldana, V. Dossetti, C. Huepe, V. M. Kenkre, and H. Larralde, "Phase transitions in systems of self-propelled agents and related network models." *Physical Review Letters*, vol. 98(9), pp. 095702, 2007. https://doi.org/10.1103/PhysRevLett.98.095702

35. H. J. Schöpe, G. Bryant, and W. Van Megen, "Effect of polydispersity on the crystallization kinetics of suspensions of colloidal hard spheres when approaching the glass transition." *The Journal of Chemical Physics*, vol. 127(8), pp. 084505, 2007. https://doi.org/10.1063/1.2760207

36. P. Dolai, A. Simha, and S. Mishra, "Phase separation in binary mixtures of active and passive particles." *Soft Matter*, vol. 14(29), pp. 6137–6145, 2018. https://doi.org/10.1039/c8sm00222c

37. E. Mones, A. Czirók, and T. Vicsek, "Anomalous segregation dynamics of self-propelled particles." *New Journal of Physics*, vol. 17(6), pp. 063013, 2015. https://doi.org/10.1088/1367-2630/17/6/063013

38. X. Yang, M. L. Manning, and M. C. Marchetti, "Aggregation and segregation of confined active particles." *Soft Matter*, vol. 10(34), pp. 6477–6484, 2014. https://doi.org/10.1039/c4sm00927d

39. A. Costanzo, J. Elgeti, T. Auth, G. Gompper, and M. Ripoll, "Motility-sorting of self-propelled particles in microchannels." *EPL (Europhysics Letters)*, vol. 107(3), pp. 36003, 2014. https://doi.org/10.1209/0295-5075/107/36003

40. N. K. Agrawal, and P. S. Mahapatra, "Alignment-mediated segregation in an active-passive mixture." *Physical Review E*, vol. 104(4), pp. 044610, 2021. https://doi.org/10.1103/PhysRevE.104.044610

41. P. S. Mahapatra, S. Mathew, M. V. Panchagnula, and S. Vedantam, "Effect of size distribution on mixing of a polydisperse wet granular material in a belt-driven enclosure." *Granular Matter*, vol. 18(2), pp. 30, 2016. https://doi.org/10.1007/s10035-016-0633-1

42. P. S. Mahapatra, and S. Mathew, "Activity-induced mixing and phase transitions of self-propelled swimmers." *Physical Review E*, vol. 99(1), pp. 012609, 2019. https://doi.org/10.1103/PhysRevE.99.012609

43. N. K. Agrawal, and P. S. Mahapatra, "Effect of particle fraction on phase transitions in an active-passive particles system." *Physical Review E*, vol. 101(4), pp. 042607, 2020. https://doi.org/10.1103/PhysRevE.101.042607

44. C. C. Liao, S. S. Hsiau, and K. To, "Granular dynamics of a slurry in a rotating drum." *Physical Review E*, vol. 82(1), pp. 010302, 2010. https://doi.org/10.1103/PhysRevE.82.010302

45. S. Mohapatra, S. Mondal, and P. S. Mahapatra, "Spatiotemporal dynamics of a self-propelled system with opposing alignment and repulsive forces." *Physical Review E*, vol. 102(4), pp. 042613, 2020. https://doi.org/10.1103/PhysRevE.102.042613

Chapter 8

A CFD model for parametric study of pipe elbow erosion behavior caused by the slurry flow

Nilesh Kumar Sharma and Satish Kumar Dewangan
National Institute of Technology, Raipur, India

Pankaj Kumar Gupta
Guru Ghasidas University, Bilaspur, India

CONTENTS

8.1 INTRODUCTION

Slurry maintained at low pressure and temperature are transported through long pipelines in many industries like oil and gas transport industry, transportation of ash and coal-water in thermal power plant, mineral processing plant, chemical, and mining unit, etc. Pipe bend and elbow are considered the most critical part of the pipeline unit in which entrainment of solid particulates deposit and continuous impacting slurry leads to material loss and causes erosion wear of the inner surface of the pipeline. The combined

DOI: 10.1201/9781003257691-8

effect of sharp pressure drop, centrifugal action at the pipe bend, and the extreme inflow region in the field of flowing fluid causes slurry deposition and solid particle impingement. Nowadays many engineering industries that are dealing with the slurry flow are looking for a lifelong solution for erosion wear of pipeline damage so that equipment maintenance cost, environmental burden, a threat to reliability and safety must be reduced to a major extent. In general for long life, it is very important to identify the location of erosion in the pipeline used for long-distance transmission.. Erosion of pipelines induced by pressurized slurry is often a dynamic phenomenon to be interpreted as it depends on several variables such as flow properties of slurry, solid characteristics, and geometric configurations of pipe bend [1–3]. Furthermore, the experimental study of erosion wear in pipe elbow is very hard to perform so the CFD method helps us to exactly study the behavior of slurry flow at each section of pipe and to identify the location of severe damage at the inner surface of the pipe elbow for further remedial action to be taken. In this research work, a computational model is used to investigate the surface erosion of the pipe elbow material and to predict the effect of various dependent parameters like solid particle concentration, particle size, and carrier fluid with the variation of slurry velocity that leads to erosion. Chen et al. [4] have studied the erosion rate behavior on the elbow and plugged tee shape pipe due to water and sand flow and also validated the result in CFD, and they also observed the erosion rate at different flow velocities. Peng et al. [5] used a CFD model to study the erosion in pipe bend and also to understand the erosion mechanism of slurry flow, the correlation between gaseous and liquid phase, solid particle motion, and erosion behavior have been studied. They have carried the investigation for the erosion in the pipe at various bend angles, and with the experimental results, they have concluded that the erosion characteristics inside the pipe wall are mainly due to the impacting particle collision phenomenon. McLaury [6] and Shirazi et al. [7] have also introduced a model for predicting erosion in elbows. Bozzini et al. [8] performed combined experimental and numerical analysis for evaluating the effect of various parameters like slurry velocity, solid particle concentration, and carrier fluid volume concentration on surface erosion of a pipe bend at different slurry flow conditions. Forder et al. [9] have developed a CFD model for erosion rate prediction at the oilfield pipeline and they have concluded that erosion wear in the pipe is directly dependent on the particle impact characteristics, pipe material property, and particle size. Most researchers have studied the erosion of pipe elbow but they lacked in terms of slurry flow behavior, and the exact measurement of erosion wear by different flow conditions still needs to be evaluated at different pipe bend locations. In the present work, CFD simulation was performed for 90° pipe elbows for parametric analysis of erosion rate at various dependent factors and to study slurry flow behavior. Erosion wear phenomenon is studied by using Euler–Lagrange model with standard $k\varepsilon$ turbulence model [7, 10]. Various pipe bend cases are studied in previous literatures [11–17]. Pipe boundary conditions are taken into account for CFD modeling at the pipe inlet, exit, and outside wall. With sand particle injection, water is permitted to enter from the pipe inlet, and the pipe is made of stainless steel 316 (SS316).

8.2 METHODOLOGY

In this section, we have considered fluid as a continuum and its motion is resolved using the Eulerian model by RANS equations while the DPM model is considered for exactly tracking the discrete phase, i.e. solid particles. Erosion rates in the pipe elbow are calculated using the recorded values from the rebound model of ANSYS FLUENT 19.0 version based on various parameters like particle impact frequency, slurry velocity, impacting particle angle, and location of impinging particle.

8.2.1 Governing equation

The slurry flow inside the pipe elbow is modeled using the ANSYS FLUENT ver. 19.0 software, Newton's second law is used to solve problems involving discrete particles, and Navier–Stokes equations are used to study fluid motion. The velocity of both phases is considered the same. The fundamental equations of mass and momentum conservation are as follows:

The mass conservation equation is given by

$$\frac{\partial \rho}{\partial t} + \nabla \cdot (\rho \vec{u}) = 0 \tag{8.1}$$

The momentum equation is given by

$$\frac{\partial}{\partial t}(\rho \vec{u}) + \nabla \cdot (\rho u_i u_j) = -\nabla p + \nabla \cdot [\mu(\nabla \vec{u} + \nabla \vec{u}^T)] + \rho g \tag{8.2}$$

8.2.2 Turbulence modeling

The turbulent kinetic energy is given by

$$\frac{\partial}{\partial t}(\rho k) + \nabla \cdot (\rho \vec{u} k) = \nabla \left[\left(\mu + \frac{\mu_t}{\sigma_k} \right) \nabla k \right] + g_k - \rho \varepsilon \tag{8.3}$$

Dissipation of turbulent KE transport equation:

$$\frac{\partial}{\partial t}(\rho \varepsilon) + \nabla \cdot (\rho \vec{u} \varepsilon) = \nabla \left[\left(\mu + \frac{\mu_t}{\sigma_\varepsilon} \right) \nabla \varepsilon \right] + \frac{\varepsilon}{k}(C_{\varepsilon_1} g_k - \rho \varepsilon C_{\varepsilon_1}) \tag{8.4}$$

where g_k is turbulence KE production average velocity gradient.

The turbulent eddy viscosity is given by

$$\mu_t = C_\mu \rho \frac{k^2}{\varepsilon} \tag{8.5}$$

Coefficients for standard $k-\varepsilon$ turbulence model is given by

$$C_\mu = 0.09, \sigma_k = 1, \sigma_\varepsilon = 1.3, C_{\varepsilon_1} = 1.44, C_{\varepsilon_1} = 1.92$$

8.2.3 Particle tracking model

As the solid particle concentration is less than 10% in this work, DPM model is used for tracking the solid particle motion in the Lagrangian frame; in this the following forces are considered:

$$m_p \frac{\partial \overrightarrow{v_p}}{\partial t} = \overrightarrow{F_D} + \overrightarrow{F_p} + \overrightarrow{F_B} + \overrightarrow{F_g} + \overrightarrow{F_{VM}} \tag{8.6}$$

The equation for Drag force is given by

$$\overrightarrow{F_D} = \frac{18\mu}{\rho_p d_p^2} \frac{C_d \operatorname{Re}_p}{24} (\vec{v} - \overrightarrow{v_p}) \tag{8.7}$$

The coefficient of drag C_d,

$$C_d = a_1 + \frac{a_2}{\operatorname{Re}_p} + \frac{a_3}{\operatorname{Re}_p{}^2} \tag{8.8}$$

where a_1, a_2, and a_3 are constants.

The pressure force $\overrightarrow{F_p}$ of the solid particle is given by

$$\overrightarrow{F_p} = \left(\frac{\rho}{\rho_p}\right)\nabla P \tag{8.9}$$

Buoyancy force and gravity force are combined for fluid is given as:

$$\overrightarrow{F_B} = \left(\frac{\rho_p - \rho_1}{\rho_p}\right)\vec{g} \tag{8.10}$$

The virtual mass force $\overrightarrow{F_{VM}}$,

$$\overrightarrow{F_{VM}} = \frac{\pi d_p^3}{12} \rho \frac{\partial(\vec{v} - \overrightarrow{v_p})}{\partial t} \tag{8.11}$$

8.2.4 Erosion equation

The erosion at the inner surface of pipe elbow caused by impinging slurry flow is predicted by the impacting solid particle properties stored at the cell next to the wall. The following equation determines erosion prediction by slurry flow in CFD code in FLUENT.

$$E_{erosion} = \sum_{p=1}^{N_{Particle}} \frac{\dot{m}_p C(d_p) f(\alpha) v^{b(v)}}{A_{face}} \tag{8.12}$$

where $\dot{m}_p$ is the mass flow rate (kg/sec), $C(d_p)$ solid particle diameter function, $f(\alpha)$ is impingement angle function (α), v is impinging velocity, $b(v)$ is a velocity factor of particles, and A_{face} is the cell face area of the cell with the wall. Once the particle trajectories have been calculated and wall impingement results have been collected, the aforementioned equation can be utilized to predict the erosion rate for individual area elements. Subsequently, to obtain the erosion rate profile for the entire geometry, this detail can be combined. The maximum erosion rates (kg·m^{-2}·sec^{-1} due to slurry flow are measured from the simulation results. Sand particles often collide with the inner wall of the pipe surface during slurry flow motion and rebound to the fluid domain.

8.3 MODEL IMPLEMENTATION PROCEDURE

8.3.1 Computational procedure

In the present study, a standard 90° elbow pipe considered to simulate the mixture of water–sand flow with particle size and five different volume fractions, i.e. 2%, 4%, 6%, 8%, and 10%, is ingested through a pipe bend made of stainless steel 316 (SS 316). The dimension of the pipe and flow conditions is shown in Figure 8.1(a) and 8.1(b). The current CFD simulation developed from FLUENT 19.0 in which FEM is used to solve the governing equation. The velocity of particle injecting from the inlet surface is equal to the carrier fluid velocity. For multiphase modeling, SIMPLE method is used for pressure-velocity coupling between fluid and solid particles. The computational model of the considered pipe elbow and all the input parameters and boundary conditions to perform simulation are mentioned in Table 8.1.

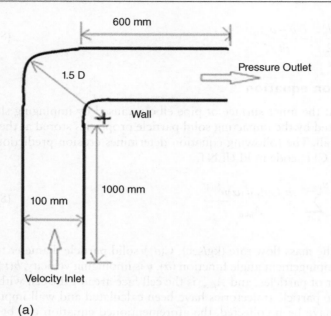

(a)

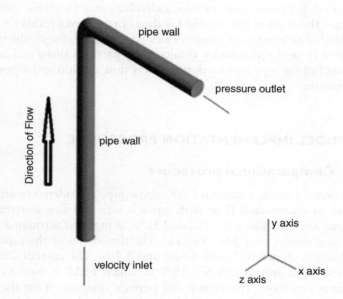

(b)

Figure 8.1 Two-dimensional diagram for pipe elbow. (a) Model geometry; (b) CFD flow domain model with boundary conditions (based on Kannojiya et al. [18]).

Table 8.1 CFD simulation parameters (data based on Kannojiya et al. [18])

Parameter	Details
Solid particle material (phase two)	Sand
Particle density	2650 kg/m^3
Particle size (μm)	50–300
Sand particle concentration (% by volume)	2%–10%
Flowing fluid (phase one)	Water (a) Density (ρ) = 997.0 kg/m^3 (b) Viscosity (μ) = 1.003 × 10^{-3} kg·m^{-1}·sec^{-1}
Elbow angle	90°
Pipe material	Stainless steel 316 (SS316)
Inlet velocity for water and sand (m/s)	2.5–15
Wall treatment	Standard wall function-no slip
Inlet flow	Velocity inlet
Outlet flow	Pressure outlet
Flow turbulence model	Standard $k-\varepsilon$ model
Pipe erosion model	Finnie model
Particle tracking	DPM model
Solver equation	Navier–Stokes(Water) and DPM (Sand)
Converge criteria	10^{-6}

8.3.2 Grid independence test

To accurately capture the erosion in the pipe elbow, a CFD model is applied and a grid independence test is performed with hexahedral and tetrahedral mesh types. We used three different sizes of mesh as fine, medium, and coarse. Simulation is performed for a slurry inlet velocity of 7.5 m/s with 100 μm particle size. The computational pipe elbow domain created using ICEM CFD and the pipe elbow consists of 986,222 number of elements, and tetrahedral mesh is used for simulation after performing grid independence test (Figure 8.2).

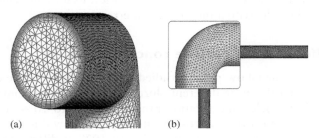

(a) (b)

Figure 8.2 Meshed geometry (tetrahedral). (a) Cross-section mesh; (b) Elbow section mesh.

Table 8.2 Grid independence test

Mesh type	Mesh resolution	Number of nodes	Number of elements	Maximum erosion rate $(kg·m^{-2}·sec^{-1})$
Hexahedral	Coarse	96,921	4,52,712	2.02E-03
	Medium	145,793	7,10,258	2.15E-03
	Fine	231,369	1,164,954	2.19E-03
Tetrahedral	Coarse	76,056	3,63,935	1.95E-03
	Medium	129,562	6,60,506	2.08E-03
	Fine	184,983	9,86,222	2.62E-03

The variation in maximum erosion rate density in the elbow pipe was less than 3%, with a maximum deviation of 6%, according to simulation data using coarse, medium, and fine grid sizes. Table 8.2 shows the maximum erosion rate for tetrahedral mesh; the difference in erosion rate between fine and medium grid size is small (less than 2%). As a result, the solution is not much sensitive to grid size. The simulation data is nearly identical to the reference experimental results and the computational time taken by hexahedral fine grid size is more than that of tetrahedral mesh with fine grid size. Hence, it is used for analysis and presented in this study.

8.4 RESULTS AND DISCUSSION

8.4.1 CFD validation

The validation of pipe elbow erosion is implemented under the same experimental condition as that of Chen et al. [4] to verify the present model. The previous researchers have investigated erosion rate in pipe bend at different configurations as elbow and plugged tee. In this validation model, the particle-fluid two-phase flows enter the 50-mm-diameter pipe bend having water–sand flow at different velocities as 15.24, 30.48, and 45.72 m/s from the vertical downward pipe and discharged fluid horizontally. From Figure 8.3, it is clear that the present simulation exactly matches with the experimental results with a minor percent error.

8.4.2 Effect of sand particle concentration

The erosion in the elbow pipe is studied under various particle concentrations. It is carried out at different slurry flow velocities by impeding the flow of solid particle concentration range from 2% to 10%, as shown in Figure 8.4. The effect on the erosion rate of pipe elbow due to variation in the sand particle concentrations from 2% to 10% at different slurry flow velocities is studied.

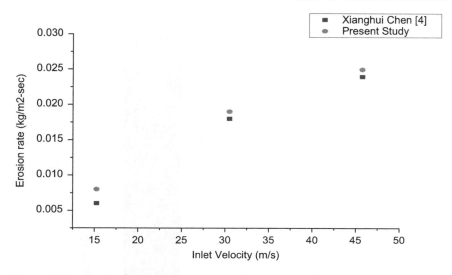

Figure 8.3 Validation of simulation results for erosion of pipe elbow.

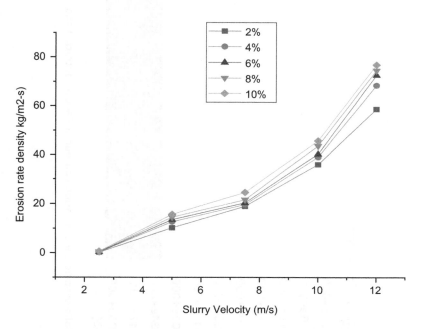

Figure 8.4 Variation of erosion at the pipe elbow for different solid particle concentration.

Figure 8.5 Solid particle concentration contour at 4% concentration and at 7.5 m/s slurry.

Erosion wear in the pipe wall is found to be increasing with the greater amount of sand in water which is represented by the simulation data shown in Figure 8.5. The quantity of particles contacting the inner surface increases as the particle concentration increases, causing severe erosion at the pipe elbow.

8.4.3 Effect of slurry velocity

Erosion of pipe elbow is investigated using the different slurry velocity, and it is clear from the literature survey that the slurry velocity is an important effect on the erosion of pipelines. Hence, we are very much concerned about the same. In this section, we have simulated the pipe flow using the slurry velocities in the range of 2.5–15 m/s at a solid particle concentration of 8% by volume. It is observed from Figure 8.6 that at higher intensity of slurry flow, high erosion rate is seen at the pipe wall due to high KE of slurry flow imparted to the wall which causes major damage to the inner surface. Figure 8.6 shows the contours for erosion at 7.5 m/s slurry velocity for 8% particle concentration and variation of erosion at pipe elbow for particular slurry velocities is shown in Figure 8.7. From the erosion profile, it can be found that erosion is mainly predominant at the elbow and bend section.

8.4.4 Effect of sand particle size

In the present simulation work, the particle size varied from 50 micron (μm) to 300 micron (μm) for slurry velocity of 7.5 m/s which is shown in

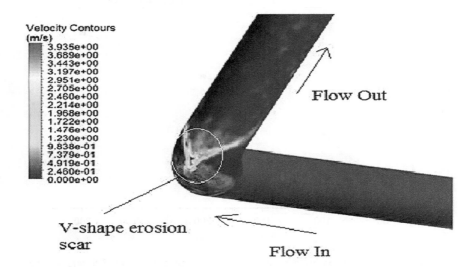

Figure 8.6 Erosion behavior at the slurry velocity of 7.5 m/s solid particle concentration at 4%.

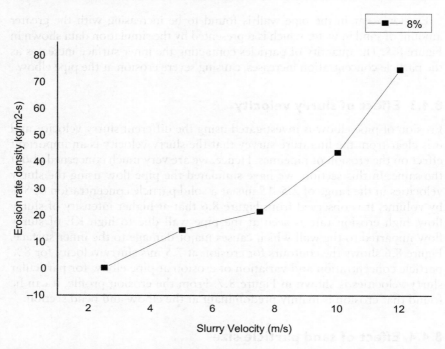

Figure 8.7 Effect of slurry velocity at 8% particle concentration for erosion rate density at pipe elbow.

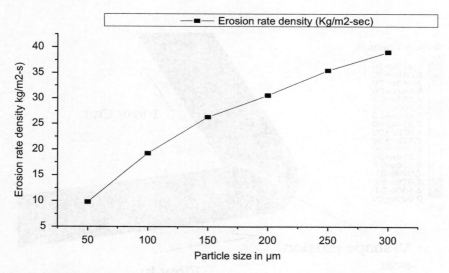

Figure 8.8 Effect on erosion rate at different particle sizes at 7.5 m/s slurry velocity.

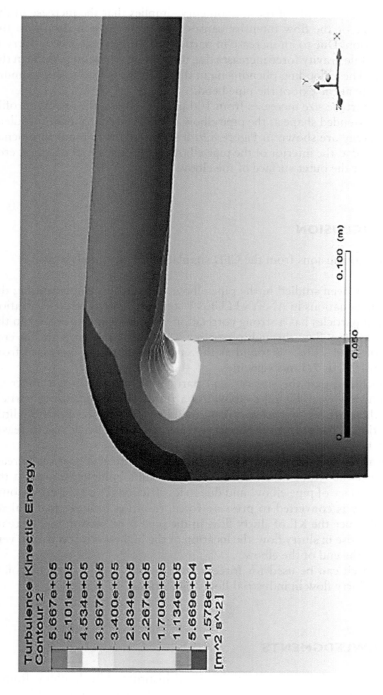

Figure 8.9 Turbulence KE at the bend section of the elbow at 7.5 m/s slurry velocity and 4% solid particle concentration.

Figure 8.8. The profile of the erosion curve implies that the increase in the particle size or the flow intensity causes maximum erosion deep into the bend position. Due to an increase in particle size, drag force in the slurry is reduced while gravity force increases due to flooding of a sand particle in the pipe flow. These flooding phenomena in the pipe elbow lead to a maximum erosion rate at the exit of the pipe bend.

As the particle size increases from 100 μm to 150 μm, the erosion profile becomes extended shape at the pipe elbow outlet. The contours of turbulent kinetic energy are shown in Figure 8.9. It shows the maximum turbulence zone created at the interior of the pipe elbow which leads to the highest erosion wear at the outer surface of the elbow.

8.5 CONCLUSION

The main conclusions from the CFD simulation of pipe analysis are:

Erosion has been studied in the pipe elbow geometry by implementing the erosion equations in ANSYS FLUENT software code. The concentration of solid particles has a strong correlation with erosion rate density in the case of slurry flow through pipe elbow. Almost three times greater erosion of bend surface is found as the sand concentration increased from 2% to 10% at 7.5 m/s velocities.

It is concluded that the size of sand particle has no significant effect on the erosion wear for pipe elbow. As a result of the increasing particle size, there are less particles striking the pipe wall. The results show little variation of increase in erosion wear when the solid particle size increases from 100 μm to 150 μm.

By increasing the inlet slurry velocity, the erosion rate also goes on increasing since the KE of flow increases which results in higher impact at the inner surface of pipe elbow, and due to less frictional loss, a large amount of energy is converted to pressure force, resulting in more material removal. Since the KE of slurry flow in the pipe bend section will increase with the rise in slurry flow, the location of the highest erosion will be very near to the end of the elbow.

This research can be used for forecasting erosion wear induced by solid–liquid slurry flow in industrial flow applications.

ACKNOWLEDGMENTS

The authors would like to thank National Institute of Technology, Raipur (C.G.), for permitting us to use its ANSYS FLUENT software and library.

NOMENCLATURE

ρ Density (kg/m³)

$\vec{u}$ Velocity vector (m/s)

ε Dissipation rate of turbulent energy (m² s⁻³)

μ Dynamic viscosity (Pa-s)

σ_k Effective Prandtl number for k

σ_ε Effective Prandtl number for ε

k Turbulence kinetic energy (m² s⁻²)

REFERENCES

1. Zhang, H., Tan, Y., Yang, D., Trias, F. X., Jiang, S., and Sheng, Y. "Numerical investigation of the location of maximum erosive wear damage in the elbow: Effect of slurry velocity, bend orientation and angle of elbow". *Powder Technology*, 217, 467–476, 2012.
2. Vieira, R. E., Mansouri, A., McLaury, B. S., and Shirazi, S. A. "Experimental and computational study of erosion in elbows due to sand particles in water-flow". *Powder Technology*, 288, 339–353, 2016.
3. Jafari, M., Mansoori, Z., Avval, M. S., and Ahmadi, G. "The effects of wall roughness on erosion rate in gas-solid turbulent annular pipe flow". *Powder Technology*, 217, 248–254, 2015.
4. Chen, X., McLaury, B. S., and Shirazi, S. A. "Application and experimental validation of a computational fluid dynamics (CFD)-based erosion prediction model in elbows and plugged tees". *Computational Fluids*, 33, 1251–1272, 2004.
5. Peng, W., and Cao, X. "Numerical prediction of erosion distributions and solid particle trajectories in elbows for gas-solid flow". *Journal of Natural Gas Science and Engineering*, 30, 455–470, 2016.
6. McLaury, B. "Predicting solid particle erosion resulting from turbulent fluctuations in oilfield geometries". Ph.D. Dissertation, the University of Tulsa, Tulsa, OK, 1996.
7. Shirazi, S., Shadley, J., McLaury, B., and Rybicki, E. "A procedure to predict solid particle erosion in elbows and tees". *Journal of Pressure Vessel Technology*, 117(1), 45–52, 1995.
8. Bozzini, B., Ricotti, M. E., Boniardi, M., and Mele, C. "Evaluation of erosion-corrosion in multiphase flow via CFD and experimental analysis". *Wear*, 255, 237–245, 2003.
9. Forder, A., Thew, M., and Harrison, D. "A numerical investigation of solid particle erosion experienced within oilfield control valves". *Wear*, 216(2), 184–193, 1997.
10. Safaei, M. R., Mahian, O., Garoosi, F., Hooman, K., Karimipour, A., and Kazi, S. N. "Investigation of micro- and nanosized particle erosion in a 90° pipe bend using a two-phase discrete phase model". *The Scientific World Journal*, 2014, 740578, 2014.

11. Vieira, R. E., Mansouri, A., McLaury, B. S., and Shirazi, S. A. "Experimental and computational study of erosion in elbows due to sand particles in water flow". *Powder Technology*, 288, 339–353, 2016.
12. Wang, K., Li, X., Wang, Y., and He, R. "Numerical investigation of the erosion behavior in elbows of petroleum pipelines". *Powder Technology*, 314, 490–499, 2017.
13. Sommerfeld, M., and Huber, N. "Experimental analysis of modeling of particle-wall collisions". *International Journal Multiphase Flow*, 25, 1457–1489, 1999.
14. Patil, M. S., Deore, E. R., Jahagirdar, R. S., and Patil, S. V.. "Study of the parameters affecting erosion wear of ductile material in solid-liquid mixture". *Proceedings of the World Congress on Engineering*. Vol III, 2011.
15. Kannojiya, V., Deshwal, M., and Deshwal, D. "Numerical analysis of solid particle erosion in pipe elbow". *Materials Today: Proceedings*, 5, 5021–5030, 2018.
16. Khur, W. S. and Jian, Y. "CFD study of sand erosion in pipeline". *Journal of Petroleum Science and Engineering*, 176, 269–278, 2019.
17. Shrestha, U., Chen, Z., Park, S. H., and Choi, Y. D. "Numerical studies on sediment erosion due to sediment characteristics in Francis hydro turbine". *IOP Conference Series: Earth and Environmental Science*, 240, 042001, 2019.
18. Peng, W., Cao, X., Hou, J., Xu, K., Fan, Y., and Xing, S. "Experiment and numerical simulation of sand particle erosion under slug flow condition in a horizontal pipe bend". *Journal of Natural Gas Science and Engineering*, 76, 103175, 2020.

Chapter 9

Dynamics of forced particles in an oscillating flow at low Reynolds numbers

Jogender Singh and C. V. Anil Kumar
Indian Institute of Space Science and Technology,
Thiruvananthapuram, India

CONTENTS

INTRODUCTION: BACKGROUND OF THE PROBLEM

The dynamics of suspended particles has significance in many engineering industries such as oil recovery and refinery, printing and papermaking, pharmaceuticals, and food processing. Suspension rheology study may lead to insights that provide better control of fluid parameters such as stress deformation and may lead to appropriate changes in processing parameters. In most situations, the dynamics of suspended particles are very sensitive to particle orientation distributions due to their irregular or non-spherical shape. It has been known to the research community since the work of Jeffrey (1922) and Bretherton (1962), especially the analysis of the problem at zero Reynolds numbers in the absence of inertia. They have found that the orbits depend on the initial conditions by neglecting the Brownian motion and the induced hydrodynamic interactions, etc. Houghton (1968) investigated the trajectories and terminal velocity of suspended particles in an oscillating fluid. The transport of particles of arbitrary shape in shear flows, where the Reynolds numbers is low, has been investigated and is summarized by Leal and Hinch (1971). Li and Sarkar (2005) have published a numerical simulation of dynamics of fluid drop in an oscillating flow in the case of a finite Reynolds number. They studied the effect of inertial-facial tension and periodic forcing at low but finite inertia. Nilsen and Andersson (2013) investigated the chaotic behavior of transport of spheroid particles considering

DOI: 10.1201/9781003257691-9

strong inertia, for the larger Stokes number. Wherein they have observed that a strange attractor is occurred for Stokes's number larger than certain critical value. The transport properties and characterization of externally driven particles are of interest to many researchers, due to their application in systems. Recently, Singh and Kumar (2019) studied the dynamics of the suspended particles in a fluid at rest in the limit low Reynolds numbers. They have discussed the properties of orientation dynamics of a spheroid suspended in an oscillating fluid at low Reynolds numbers. The properties are characterized with respect to the change in parameter values such as aspect ratio, particle-fluid density, natural frequency, and amplitude of driving force. In this work, the governing equation of transport of a periodically forced body suspended in an oscillating Newtonian fluid in the limit of low Reynolds number is derived and then the solutions of the nonlinear differential equations are numerically studied in detail, following the formulation of Lovalenti and Brady (1993) and the second law of motion.

FORMULATION AND METHODOLOGY

The general expression of the hydrodynamic force exerted on a solid particle suspended in a time-dependent Newtonian fluid is developed by Lovalenti and Brady (1993) as:

$$
F_s^H = ReSl\, \tilde{V}_p\, u_\infty(t) + F_s^H(t) - ReSl
$$

$$
\left[6\pi \cdot \Phi \cdot \Phi \cdot \Phi + \lim_{R \to \infty} \left(\int_{V_f(R)} M^T \cdot M \, dV - \frac{9\pi}{2} \Phi \cdot \Phi R \right) \right] \cdot \dot{u}_s(t)
$$

$$
+ \frac{3}{8} \left(\frac{ReSl}{\pi} \right)^{\frac{1}{2}} \int_{-\infty}^{t} \left\{ \frac{3}{2} F_s^{H\parallel}(t) - \left[\frac{1}{|A|^2} \left(\frac{\pi^{\frac{1}{2}}}{2|A|} \operatorname{erf}(|A|) - \exp(-|A|^2) \right) \right] \right.
$$

$$
F_s^{H\parallel}(s) + \frac{2}{3} F_s^{H\perp}(t) - [\exp(-|A|^2)
$$

$$
\left. - \frac{1}{|A|^2} \left(\frac{\pi^{\frac{1}{2}}}{2|A|} \operatorname{erf}(|A|) - \exp(-|A|^2) \right) \right] F_s^{H\perp}(s) \right\} \frac{2ds}{(t-s)^{\frac{3}{2}}} \cdot \Phi
$$

$$
- Re \lim_{R \to \infty} \int_{V_f(R)} (u_0 \cdot \nabla u_0 - u_s(t) \cdot \times \nabla u_0) \cdot M \, dV + O(ReSl) + O(Re)
$$

Here,

$$
Y(s) = Y_s(t) - Y_\infty(s).
$$

We assume the velocity of the fluid as $u_\infty(t) = (u_\infty(t), v_\infty(t), w_\infty(t))$. Let the velocity of the particle exerted by the fluid be in the direction of the vector denoted by $\mathbf{A}$, where $\mathbf{A}$ is parallel to the displacement vector; $Y_s(t) - Y_s(s)$. Assume that the velocity component is given by $u_p(t) = (u_p(t), v_p(t), w_p(t))$. Therefore, the slip velocity of the fluid is given by $u_s(t) = u_p(t) - u_\infty(t)$. In this analysis, velocity $u_s(t)$, $u_p(t)$ and $u_\infty(t)$ is non-dimensionalized by the characteristic velocity U_c and the accelerations by U_c/τ_c, and length by a. Here, a, U_c, and τ_c are the characteristic particle parameters of the suspended particle, and characteristic timescale is defined as $\tau_c = a/U_c$. As explained in Singh and Kumar (2019), the Stokes resistance tensor associated with the particle is $6\pi\mathbf{\Phi}$, and hence the hydrodynamic force, known as the pseudo-steady state drag force, acting on the suspended particle translating with slip velocity us is given by $\mathbf{F}_s^H(t) = -6\pi\mathbf{\Phi} \cdot u_s(t)$. The second-order tensor, $\mathbf{\Phi}$ is taken as $\mathbf{\Phi} = \dfrac{8e}{3}S$, where e is the eccentricity of the spheroid and S, is a diagonal matrix defined by

$$S_{11} = \frac{e^2}{-2e + (1 + e^2)log\left(\dfrac{1+e}{1-e}\right)}$$

$$S_{22} = \frac{-2e^2}{-2e + (1 - 3e^2)log\left(\dfrac{1+e}{1-e}\right)}$$

$$S_{33} = S_{22},$$

and

$$S_{ij} = 0 \text{ for } i \neq j.$$

The first term in the above expression, $\mathbf{F}^H(t)$ is due to an accelerating frame of reference translating with the particle, the second term is due to *the pseudo-steady Stokes drag* (Stokes's force). The third one is *the unsteady Oseen correction* due to the hydrodynamic force. In fact, this is a history integral term that replaces Basset memory contribution, and the fourth term is *an acceleration reaction term*. The last integral term is the contribution of a force orthogonal to the direction of particle velocity, named *the lift force*. Note that pseudo-Stokes force term $\mathbf{F}_s^H$ can be conventionally split into, namely parallel and perpendicular components to the displacement vector. Thus, the parallel and perpendicular components of the pseudo-Stokes force is respectively given by

$$\mathbf{F}_s^{H\parallel} = 6\pi\, u_s \cdot \mathbf{pp} \text{ and } \mathbf{F}_s^{H\perp} = 6\pi\, u_s \cdot (\delta - \mathbf{pp})$$

where δ is the second-order idem tensor and $\mathbf{p}$ is the normalized vector given by

$$\mathbf{p} = \frac{\mathbf{Y}_s(t) - \mathbf{Y}_s(s)}{|\,\mathbf{Y}_s(t) - \mathbf{Y}_s(s)\,|}$$

In the above term, $\mathbf{Y}_s(t) - \mathbf{Y}_s(s)$ is the displacement between where the particle starts at time, s and where the particle is now at time t, called the integrated displacement of the spheroid and $\mathbf{A}$ has been defined by

$$\mathbf{A} = \frac{\mathrm{Re}}{2}\left(\frac{ReSl}{t-s}\right)^{\frac{1}{2}}\left(\frac{\mathbf{Y}_s(t) - \mathbf{Y}_s(s)}{t-s}\right)$$

as can be seen from Singh and Kumar (2019). We consider transport of a spheroid particle in a fluid with uniform velocity which is assumed to be along the major (polar radius) axis of the spheroid. Let a and b be the semi-major (polar radius) and semi-minor (equatorial radius) axes of the spheroid and (x, y, z) represent any point on the surface of the spheroid for which the equation of the three-dimensional spheroid takes the form

$$\frac{x^2}{a^2} + \frac{y^2 + z^2}{b^2} = 1.$$

This form proposes three cases for the problem, say $a < b$ for oblate spheroid and $a > b$ for prolate spheroid; and the case of $a = b$ for a sphere. The aspect ratio α is defined as $\alpha = a/b$. In this work, we consider prolate spheroid, i.e, α is greater than 1. Note that eccentricity (e) of the prolate is given by $e = \sqrt{\dfrac{a^2 - b^2}{a^2}}$. Remember that the particle is suspended in a fluid and is driven by an external periodic force. Therefore, the dominant forces acting on the particle affecting the dynamics of the particle could be (1) a downward force due to gravity, (2) an upward buoyant force, (3) the hydrodynamic force due to disturbance of the fluid in vicinity of the particle, (4) the lift force, and (5) the applied external force.

THE EQUATIONS GOVERNING THE TRANSPORT

Upon manipulation by incorporating the above assumptions and simplifications, the expression for the hydrodynamic force exerted by the fluid on the particle reduces to

$$F^H(t) = ReSl(\tilde{V}_p \, \dot{u}_\infty(t) + (I_{xx}, I_{yy}, I_{zz}) \cdot u_\infty(t)) - 16\pi \, e(e_1 u_s(t),$$

$$e_2 v_s(s), e_2 w_s(t)) - ReSl(I_{xx} u_p(t), I_{yy} v_p(t), I_{zz} w_p(t))$$

$$+ \frac{3}{8}\left(\frac{ReSl}{pi}\right)^{\frac{1}{2}} \left\{ \frac{1024}{9} \pi \, e^2 (e_1^2 \, u_s(t), e_2^2 \, v_s(t), e_2^2 \, w_s(t)) \left(\frac{1}{\sqrt{t}} - \frac{1}{\sqrt{\varepsilon}}\right)\right.$$

$$+ \frac{256}{3} \pi e^2 \int_0^{t-\epsilon} B(e_1^2 u_s(s), e_2^2 \, v_s(s), e_2^2 \, w_s(s)) \frac{2ds}{(t-s)^{\frac{3}{2}}} \right\} - Re(L_1, L_2, L_3)(t),$$

where $L_1(t)$, $L_2(t)$, $L_3(t)$ are the components of the lift force obtained from

$$\lim_{R \to \infty} \int_{V_f} (u_0 \cdot \nabla u_0 - u_s(t) \cdot \nabla u_0) \cdot M \, dV$$

We need to find out expression for u_0 and $u_s(t)$, where $u_s(t)$ is the slip velocity of the particle given by the expression, $u_s(t) = u_p(t) - u_\infty(t)$. Note that M is known to us and can be calculated as given in Singh and Kumar (2019), whereas the expression for B is given by

$$B = \frac{1}{|A|^2}\left(\frac{\pi^{\frac{1}{2}}}{2|A|} \text{erf}(|A|) - \exp(-|A|^2)\right).$$

In the present analysis, the external force F^{ext} is assumed as dimensionless periodic force and is taken as $F^{ext} = (F_1 \sin(t); F_2 \sin(t); F_2 \sin(t))$, where time has been scaled with $\tau_c \left(= \dfrac{a}{U_c}\right)$. The characteristic velocity U_c is assumed to be $a\omega_1$, and ω_1 is the frequency of external periodic force. Using the above expression for F^H the governing equations of the motion of a prolate spheroid, starting with zero velocity at $t = 0$ is obtained. The slip velocity of the particles is taken as $u_s(t) = (u_p(t), v_p(t), 0) - (u_\infty \sin(\omega t), v_\infty \sin(\omega t), 0)$ which is scaled by the length of the particle and the external force frequency, i.e. by $U_c = a\omega_1$. Let $Y_p(t) = (x(t), y(t), z(t))$ be the displacement vector, and $u_p(t)$ is the velocity vector of the particle. Let F^{ext} denotes the external force acting on the particle, and hence the total force applied on the particle is $F^H(t) + F^{ext}(t)$. Now, the governing equation using Newton's law of motion is given by

$$m_p \frac{d}{dt} u_p(t) = F^H(t) + F^{ext}(t)$$

where m_p is the mass of the particle. In dimensionless form, the above equation reduces to

$$\frac{m_p \dot{u}_p}{\mu a^2 / U_c} = F^H(t) + F^{ext}(t)$$

Now, the set of equations governing the dynamics can be made available using $\frac{dYp(t)}{dt} = u_p(t)$, along with the above equation. The lift force for the prolate spheroid is numerically computed as per the scheme outlined in Singh and Kumar (2019). From the numerical simulations, we have found that the lift force is very small, i.e. $O(10^{-8})$, in this case and the reason could be that it acts only along the perpendicular direction of the force. Otherwise also, the external force is applied along the direction of the motion of the spheroid and hence the lift force is zero. However, the term is ignored for further analysis, since it is very small as per the computation. The set of integro-differential equations, thus obtained are as follows:

$$\frac{du_p(t)}{dt} = \frac{1}{C_1 \, Re} \left[F_1 \sin(t) + ReSl \left(\frac{4}{3}\pi a^2 + I_{xx} \right) u_\infty \, \omega \cos(\omega t) \right.$$

$$\left. -16\pi e e_1 u_s(t) - \frac{3}{8} \left(\frac{ReSl}{\pi} \right)^{\frac{1}{2}} (P1 + Q1) - Re \, L_1 \right]$$

$$\frac{dx(t)}{dt} = u_p(t)$$

$$\frac{dv_p(t)}{dt} = \frac{1}{C_1 \, Re} \left[F_2 \sin(t) + ReSl \left(\frac{4}{3}\pi a^2 + I_{yy} \right) u_\infty \, \omega \cos(\omega t) \right.$$

$$\left. -16\pi e e_2 u_s(t) - \frac{3}{8} \left(\frac{ReSl}{\pi} \right)^{\frac{1}{2}} (P_2 + Q_1) - Re \, L_2 \right]$$

$$\frac{dy(t)}{dt} = v_p(t).$$

Where,

$$C1 = \frac{4\pi}{3} \left(\frac{a}{b} \right)^2 + Sl \, I_{xx}$$

$$C2 = \frac{4\pi}{3}\left(\frac{a}{b}\right)^2 + Sl\, I_{yy}$$

$$P_1 = \frac{256\pi e^2}{3}\int_0^{t-\epsilon} B\,\frac{e^2 u_s(s)}{(t-s)^{3/2}}\,ds$$

$$P_2 = \frac{256\pi e^2}{3}\int_0^{t-\epsilon} B\,\frac{e^2 v_s(s)}{(t-s)^{3/2}}\,ds$$

$$Q_1 = \frac{1024}{9}\pi e^2 e_1^2\, u_s(t)\left(\frac{1}{\sqrt{t}} - \frac{1}{\sqrt{\varepsilon}}\right)$$

$$Q_2 = \frac{1024}{9}\pi e^2 e_2^2\, v_s(t)\left(\frac{1}{\sqrt{t}} - \frac{1}{\sqrt{\varepsilon}}\right)$$

The acceleration-reaction term (I_{xx}, I_{yy}, I_{zz}) is calculated using the expression proposed in the work of Pozrikidis (1992) and Chwang (1975). The scheme of computation is also explained in Singh and Kumar (2019). After calculating all required expressions, we proceed to find the solution of the more general set of equations given above. The displacement, velocity at time (t) are computed for different parameters like aspect ratio, Reynolds number, external force, etc., and analyzed them by constructing phase space plots.

NUMERICAL AND SIMULATION PROCEDURE

The integro-differential equations are solved numerically by implementing the trapezoidal. The set of 20000 data points of both position and velocity are computed from the governing equations in MATLAB software with time step of 0.0001. Further changes in time step and resolution did not yield any significant changes in the results in sample run of the code, and hence we kept the same values for all computations in this analysis. All possible plots are taken and the results obtained are analyzed meticulously. The perturbed solutions of the system are obtained and the same is validated in this work before proceeding to further analysis. In order to obtain the perturbed solution, we use Taylor series expansions for nonlinear integral terms of the governing equations. All expressions for arbitrary-shaped spheroid have been considered up to the order of $ReSl$. In this analysis, we select Reynolds number Re as the perturbation parameter, by choosing $Sl = 1$. The expression for hydrodynamic force induced on a solid spheroid of arbitrary shape given by Lovalenti and Brady [8] is also validated up to $O(\mathrm{Re})$. The perturbed solutions of the system are compared with our numerical solutions. The perturbed (for more details see [11]) solutions are given below:

x-component of the velocity, u_{per}:

$$u_{per}(t) = u_0 + u_1 Re^{1\backslash 2} + u_2\, Re + o(\mathrm{Re})$$

with

$$u_0 = F_1 \frac{\sin(t)}{16\pi ee_1}$$

$$u_1 = \frac{8}{3}\left(\frac{Sl}{\pi}\right)^{\frac{1}{2}} u_0 \left(\frac{1}{\sqrt{t}} - \frac{1}{\sqrt{\varepsilon}}\right) + \frac{F_1 B}{8\pi}\left(\frac{Sl}{\pi}\right)^{\frac{1}{2}} \int_0^{t-\epsilon} \frac{\sin(s)}{(t-s)^{1/2}}\,ds$$

$$u_2 = \frac{Sl\omega\, u_\infty \cos(\omega t)}{16\pi ee_1}\left(\frac{4\pi}{3}\alpha^2 + I_{xx}\right) - \frac{L_1}{16\pi ee_1} - \frac{C_1 F_1}{256\pi^2 e^2 e_1^2}$$
$$+ \frac{8}{3} ee_1 \left(\frac{Sl}{\pi}\right)^{\frac{1}{2}}\left(\frac{1}{\sqrt{t}} - \frac{1}{\sqrt{\varepsilon}}\right) u_1 + 2 Bee_1 \int_0^{t-\epsilon} \frac{u_1(s)}{(t-s)^{3/2}}\,ds$$

y-component of velocity, v_{per}:

$$v_{per}(t) = v_0 + v_1 Re^{1\backslash 2} + v_2\, Re + o(\mathrm{Re})$$

with

$$v_0 = F_2 \frac{\sin(t)}{16\pi ee_2}$$

$$v_1 = \frac{8}{3}\left(\frac{Sl}{\pi}\right)^{\frac{1}{2}} v_0 \left(\frac{1}{\sqrt{t}} - \frac{1}{\sqrt{\varepsilon}}\right) + \frac{F_2 B}{8\pi}\left(\frac{Sl}{\pi}\right)^{\frac{1}{2}} \int_0^{t-\epsilon} \frac{\sin(s)}{(t-s)^{1/2}}\,ds$$

$$v_2 = \frac{Sl\omega\, u_\infty \cos(\omega t)}{16\pi ee_2}\left(\frac{4\pi}{3}\alpha^2 + I_{yy}\right) - \frac{L_2}{16\pi ee_2} - \frac{C_1 F_1}{256\pi^2 e^2 e_2^2}$$
$$+ \frac{8}{3} ee_2 \left(\frac{Sl}{\pi}\right)^{\frac{1}{2}}\left(\frac{1}{\sqrt{t}} - \frac{1}{\sqrt{\varepsilon}}\right) v_1 + 2 Bee_2 \int_0^{t-\epsilon} \frac{v_1(s)}{(t-s)^{3/2}}\,ds$$

RESULTS AND DISCUSSION

A significant increase in the amplitude of position and a slight increase in the amplitude of velocity are observed due to the aspect ratio, α as evident from the plots given in **Figures 9.1** and **9.2** for the aspect ratios $\alpha = 2, 4, 6, 8$, and 10, $R_F = 0.5$, $Re = 0.1$, and $\omega = 0.2$. The figures depict the influence of aspect ratio on the dynamics of forced suspensions. It is noted that the increase in Reynolds number, Re of particle causes increment in the amplitude of the oscillations of the spheroid.

Plots in **Figure 9.3** show that there is a change in the constructed phase space due to the different choices of the parameters.

The effect of Reynolds number on the dynamics properties of the spheroid is investigated for different Reynolds numbers ranging from 0.02 to 0.1 in steps of 0.02. The variation of amplitude and phase of oscillations of position for different Reynolds numbers and external forces for an aspect ratio, $\alpha = 6$ and for frequency of fluid oscillations, $\omega = 0.2$ are graphed in **Figures 9.4** and **9.5**, respectively.

The results show that the amplitudes of position for prolate spheroid considered in this analysis slightly increase as the Reynolds number increases. The influence of the parameters on phase changes of particle motion becomes negligible as the Reynolds number and the amplitude of external force increases and attains a regular orbit as depicted in **Figure 9.5**.

CONCLUSIONS

In this chapter, a two-dimensional numerical analysis is carried out to determine the dynamical characteristics of a prolate spheroid suspended in an oscillating fluid at a low Reynolds number. The influence of the parameters namely, Reynolds number, aspect ratio, amplitude of external force on the dynamics of the particle is investigated and analyzed in detail. The motion under the action of a periodic force in the limit of low-Reynolds number is considered. The governing equations forming a system of integro-differential equations are derived from the Newton's second law of motion using the expression developed by Lovalenti and Brady (1993) for the hydrodynamics force. The solutions are numerically computed and analyzed. We observe that the amplitude of the displacement and velocity of the driven spheroid increase as the aspect ratio or/and frequency or/and force increase.

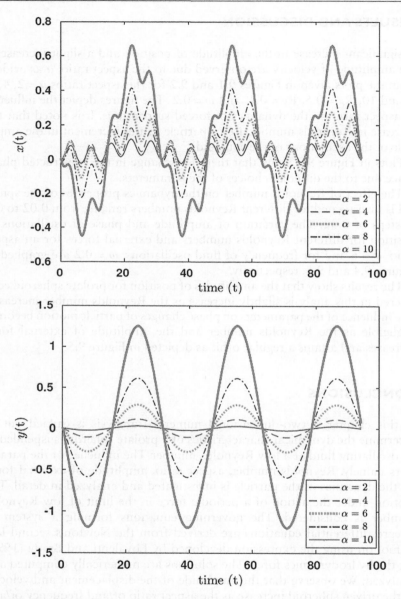

Figure 9.1 The plots showing the variation of time series of x and y components of position for different values of the aspect ratio, where Re = 0.1, $R_F = 0.5$, and = 0.2.

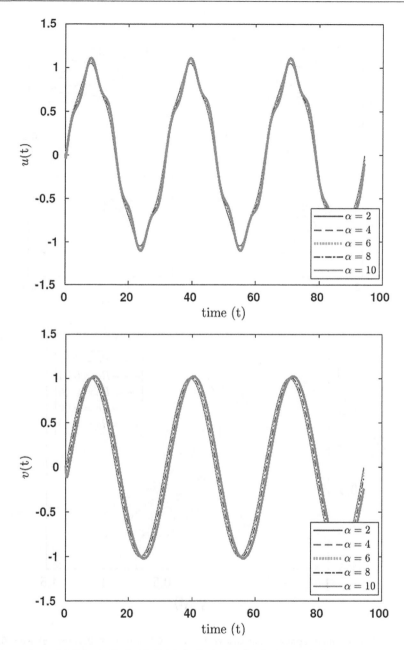

Figure 9.2 Plots showing the variation of *x* and *y* components time series of velocity for different values of the aspect ratio, when Re = 0.1, R_F = 0.5, and
$\omega = 0.2$.

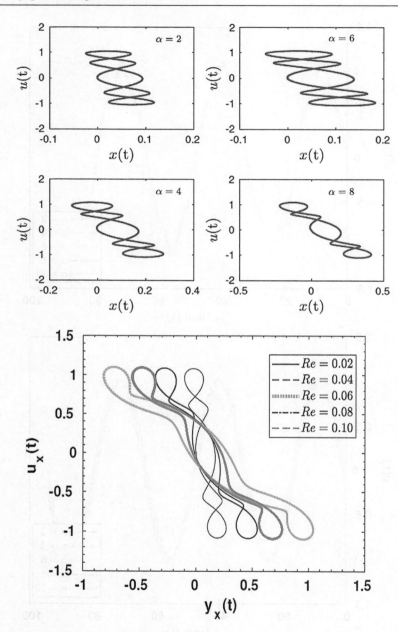

Figure 9.3 The phase space plots drawn for $R_F = 0.5$ and $\omega = 0.2$, for (a) Re = 0.1, α = 2, 4, 6, and 8 (b) Re = 0.02, 0.04, 0.06, 0.08, and 0.10, and $\alpha = 6$.

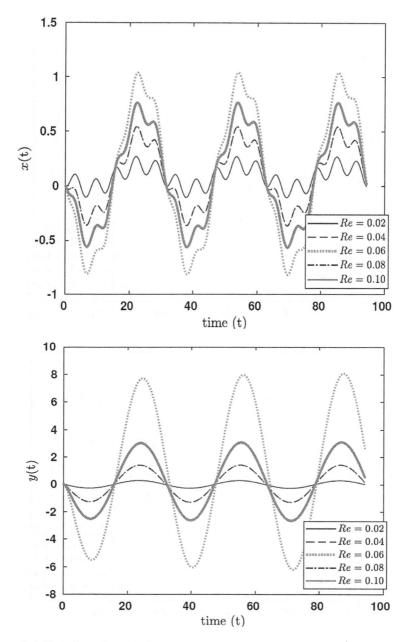

Figure 9.4 The plots showing the variation of *x* and *y* components of time series of position for different values of Re, where $\alpha = 6$, $R_F = 0.5$, and $\omega = 0.2$.

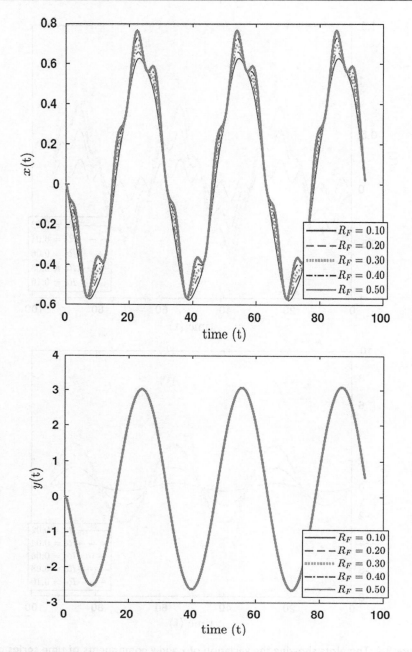

Figure 9.5 The plots showing the variation of *x* and *y* components of time series of position for different values of R_F, where $\alpha = 6$, $R_e = 0.1$, and $\omega = 0.2$.

REFERENCES

Bretherton, Francis P. 1962. "The Motion of Rigid Particles in a Shear Flow at Low Reynolds Number." *Journal of Fluid Mechanics* 14 (2): 284–304.

Chwang, Allen T. 1975. "Hydromechanics of Low-Reynolds-Number Flow. Part 3. Motion of a Spheroidal Particle in Quadratic Flows." *Journal of Fluid Mechanics* 72 (1): 17–34. https://doi.org/10.1017/S0022112075002911.

Houghton, Gerald. 1968. "Particle Retardation in Vertically Oscillating Fluids." *The Canadian Journal of Chemical Engineering* 46 (2): 79–81.

Jeffrey, George B. 1922. "The Motion of Ellipsoidal Particles Immersed in a Viscous Fluid." *Proceedings of the Royal Society of London. Series A* 102 (715): 161–79.

Leal, L. G., and E. J. Hinch. 1971. "The Effect of Weak Brownian Rotations on Particles in Shear Flow." *Journal of Fluid Mechanics* 46 (4): 685–703.

Li, Xiaoyi, and Kausik Sarkar. 2005. "Drop Dynamics in an Oscillating Extensional Flow at Finite Reynolds Numbers." *Physics of Fluids* 17 (2): 27103. https://doi.org/10.1063/1.1844471.

Lovalenti, Phillip M., and John F. Brady. 1993. "The Hydrodynamic Force on a Rigid Particle Undergoing Arbitrary Time-Dependent Motion at Small Reynolds Number." *Journal of Fluid Mechanics* 256 (29): 561–605. https://doi.org/10.1017/S0022112093002885.

Nilsen, Christopher, and Helge I. Andersson. 2013. "Chaotic Rotation of Inertial Spheroids in Oscillating Shear Flow." *Physics of Fluids* 25 (1): 13303. https://doi.org/10.1063/1.4789376.

Pozrikidis, C. 1992. *Boundary Integral and Singularity Methods for Linearized Viscous Flow.* Cambridge, UK: Cambridge University Press. https://doi.org/10.1017/CBO9780511624124

Singh, Jogender, and C.V.A. Kumar. 2019. "Dynamics of a Periodically Forced Spheroid in a Quiescent Fluid in the Limit of Low Reynolds Numbers." *Rheologica Acta* 58 (11–12). https://doi.org/10.1007/s00397-019-01169-5.

Chapter 10

Numerical study of *F1* car rear wing assembly

Manoj Kumar Gopaliya

The NorthCap University, Gurugram, India

CONTENTS

10.1 INTRODUCTION

A Formula One (*F1*) car is a compact racing car with an open cockpit and open wheel arrangement along with single-seat seating. This car with front and rear wings, and an engine placed behind the driver, designed specifically for Formula One racing events. The prevailing guidelines for these cars are exclusive to the contest and lay down that cars must be built by the participating squads only, yet the design plan and manufacturing can be subcontracted. Every *F1* car is adept at going up to 160 km/h (up to 99 mph) and back to halt in under five seconds. *F1* cars are having exceptional cornering ability apart from being reckless in straight lines. Grand Prix cars can manage corners of the racing tracks at considerably higher speeds owing to the strong ground grip and downforce. Here, aerodynamics becomes a key to winning, and teams, therefore, employ a huge amount of *R & D* activities for improving flow over these cars. The flow over these cars has two crucial apprehensions, one is the formation of downforce for better tire grip on the track along with better cornering forces, and the other is to minimize the drag due to flow separation leading to slowing down of the car.

DOI: 10.1201/9781003257691-10

10.2 LITERATURE REVIEW AND OBJECTIVE

The outcome of the literature review is utilized to get a basic understanding of some of the basic techniques and concepts used in computational fluid dynamics (CFD) analysis. Computational fluid dynamics is a branch of applied fluid dynamics that uses numerical methods to solve fluid flow problems of varying complexities. To successfully run *CFD* simulations, the basic requirements include a *CAD* model, meshing of the model, boundary conditions, and simulation testing setup. This setup helps the designer to save a lot of time and effort requirement for the physical testing of models. It provides the much-needed cost and time reduction for any problem solving with good achievable accuracy. The selection of airfoil geometry and orientation is based on an available set of papers and journals which are already validated using available test data [1, 2]. For the upper wing geometry, *NACA 2415* and lower wing geometry of *NACA 2412* was finalized. The size of the domain is selected based on the available reference materials [3]. This domain size was accepted as the thumb rule for external flow *CFD* analysis. The mesh independence test is done to obtain a solution that is free from the quality and size of the mesh. The use of k-ω-SST turbulence modeling was also selected basing results from Azmi et al. [4]. It is a good model that captures turbulence behavior at adverse pressure gradients.

The goal of the study is to analyze the behavior of a race car rear wing through a parametric study. In the analysis of the race car rear wing, there are two major quantities of importance: the amount of downforce created and the amount of drag force and the *AOA* for maximum downforce against minimum drag force. As the rear wing assembly consists of a set of two airfoils, the study also focuses on the regressive effect of the auxiliary wing on the main wing. We will study the incremental or decremented effect on parameters like drag force, drag co-efficient, and the range of allowable orientation of angle of attack. We also aim to study the parameters of airfoil during the *DRS* and Non-*DRS* zone. To simulate this scenario, the simulation will be done with the varying magnitude of air velocity corresponding to the real-life on-track situation. The geometry setups of the wing assembly during *DRS* and Non-DRS zones have always remained a non-disclosed matter from the organizations involved in *F1* racing. The study aims to estimate the optimized *AOA* for *F1* cars under all speed ranges.

10.3 MATERIALS AND METHODS

The airfoil modeling was done using Autodesk Inventor which is a *CAD* modeling software having varied uses. To generate the airfoil, we used the online application that generates the coordinates of points which when joined generates the shape of the required airfoil. The application has been referred to in the references section. All the coordinates are exported to a spreadsheet, and this is linked to the *CAD* software which generates the listed coordinates. After the coordinates appear, we use a spline curve to join the points in an orderly manner and generate the airfoil shape.

After completely constraining the sketch, we will extrude the face to generate the 3D airfoil. As the rear wing assembly demands the setup of two airfoils, both the airfoils namely NACA 2412 and NACA 2415 were created in an above-mentioned way, and then they were assembled using assemble parts feature in inventor.

Before assembling the airfoils to create the rear wing, the constraints were applied between proper planes, axes, and faces to obtain the correct AOA setup and orientation regarding gaps and spaces between them. As per the discussions above, it was fixed to keep the NACA 2415 as the main airfoil kept on the lower side and the top side NACA 2412, the thinner airfoil was placed. For varying the gap and angle of setup, some face distance constraints and plane angle constraints were provided to alter the geometry as part of the varied angle of attacks set up for different test cases (Figure 10.1).

ANSYS software package offers various features in which CAD modeling and setup unit is also provided. It is called a DESIGN MODELER. Here the assembly file created above is imported and further activities are conducted to make the geometry ready for CFD simulations.

The first and foremost thing which is done here is to covert the geometry from a 3D element to a 2D element. For this, first of all, we create an outline that serves as the flow domain of our test case. Then the surface is generated on the face of the airfoil which is on the plane of the flow domain. A surface is also created at the flow domain sketch.

After this, by using the Boolean feature of the design modeler workbench, we subtract the airfoil face surface from the flow domain surface to create a blank space in the later surface. At last, by using the body delete feature, we delete the 3D airfoil elements from our setup permanently which finally converts our geometry ready for 2D meshing purposes.

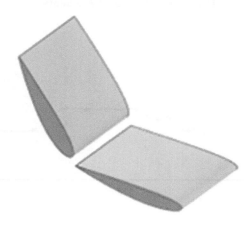

Figure 10.1 The 3-D model of the rear wing assembly.

For this model, the meshing setup includes an element size of 12 mm average, near the airfoil walls the element size is selected to be 2 mm, adaptive sizing is turned on. This feature of adaptive sizing allows a smooth transition and blending from bigger to smaller element sizes. Fiffteen layers of inflation are applied at a default growth rate of 2. The element shape is a triangle in nature and it has been justified in all the previous papers.

The mesh statistics can be monitored to keep the quality of mesh well within the acceptable limits of operability. The key mesh statistics include aspect ratio, skewness of matrix, and orthogonal property of the matrix. The acceptable values of all these parameters are well defined as far as theoretical analysis is concerned, and there is also a set of a modified set of parameters that are based on experience and hands-on performed by other scholars and research associates.

The size of the mesh determines the accuracy of the simulation results, but it increases the total computation time for the solution. Mesh independence is checked every time before starting the simulation to decrease the computation time. The mesh resizing is repeated to get the finer mesh, at a point when two or more mesh deliver nearly the same results of simulation, the coarser mesh is approved and it is used afterward (Figure 10.2).

After the finalization of the geometry and creation of acceptable mesh is complete, the next step is taken to configure the solver package. Here fluent solver is used to run the test case simulations. The setup has various aspects to it which include general setup, model selection, material designation, setting the boundary condition, choosing solution method, control parameters, report definition, monitoring, residual values, initialization, and finally running the calculation.

In general, basic setups invove the selection of pressure or density-based solver, type velocity used, and 2D space definition. In axisymmetric 2D space, the model becomes symmetric along the vertical Y-axis, and the result

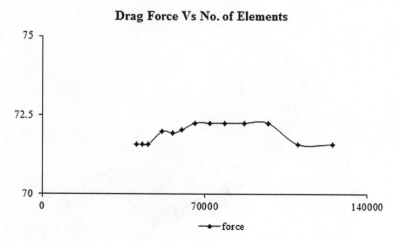

Figure 10.2 Drag force vs no. of elements for mesh independency.

is integrated on it, in planar space, the Z-component of all parameters is restricted from taking part in the simulation purpose. The axisymmetric swirl allows the user to study the effect of Z component parameters.

In the model selection menu of fluent, it allows the user to select the specific parameter study that is to be performed; it allows us to make the selection of the turbulence model that we need to use for this simulation purpose.

In the material designation part, the software allows the user to specify the materials of fluid and solids in the specified simulation. The properties of solid and fluid like the density, viscosity can be specified in this section of the software package. For our simulation, all the default values of the solid aluminum and air were kept unchanged.

Boundary setup conditions are provided as the constraints based on which the solver runs the process. Boundary conditions are constraints necessary for the solution of a boundary value problem. A boundary value problem is a differential equation (or system of differential equations) to be solved in a domain on whose boundaries set of conditions are known. In the software, boundaries can be defined by using various options available such as inlet, outlet, symmetric walls, walls with roughness, etc.

In the flow domain analyzed during the present study, there are namely four boundaries velocity inlet, pressure outlet, airfoil walls, and far walls. The velocity inlet is defined with the flow velocity of air around the model. The pressure outlet is defined with a pressure of 0 atm to achieve proper flow direction and convergence of the solutions. Both far walls and airfoil walls are kept at wall condition with all parameters set to default.

The method and control section allows us to select the various schemes and control parameters that will control the calculation run and iterate the required results.

Report definition is the most important segment in this package. It is used to mention the type of requirement result that is needed to be extracted from the calculation of the solver. Here we have defined two report definitions: one for calculating the drag force and the lift force in the simulation. The initialization and run calculation tab allows us to initialize the problem. It is performed to provide an initial guessed value of the required result to each cell, and it helps us to achieve faster convergence of the solution. The run calculation allows us to select the number of iterations and reporting intervals of the residual values.

10.4 GOVERNING EQUATIONS

A drag force in any external flow is the opposing thrust produced due to flow separation during the fluid flow over a body. This force acts reverse to the free stream velocity. The drag force *(D)* experienced by a body while traveling through a fluid is specified by:

$$D = \frac{1}{2}\rho a v^2 C_d \tag{10.1}$$

where
D is the drag force;
ρ is the fluid density;
v is the flow velocity;
a is the cross-sectional area of the body projected transverse to the direction of flow;
C_d is the coefficient of drag.

Downforce is the complementary force of the drag force which originates from moving air over the body of a vehicle. It is simply the lift generated by aerofoil in a reversed direction tending to keep the vehicle stuck to the ground. Downforce, or negative lift, pushes the car onto the track.

The expression of downforce is as same as that of the lift force (L) which is given by:

$$L = \frac{1}{2}\rho a v^2 C_L \qquad (10.2)$$

where
L is the lift force;
ρ is the air density;
v is the flow velocity;
a is the cross-sectional area of the body projected transverse to the direction of flow;
C_L is the coefficient of lift.

The non-dimensional number, Reynolds number (Re), is employed during this CFD modeling project. It is defined as the ratio of the inertia and viscous forces and measures their significance for the arranged flow condition:

$$Re = \frac{\rho U L}{\mu} \qquad (10.3)$$

where L and U are the representative length and velocity scales of the flow domain; ρ and μ are the fluid density and its dynamic viscosity, respectively. The use of the Reynolds number is commonly done while executing a dimensional analysis, known as the Reynolds Similarity Principle.

The governing equations used in the current CFD project are Continuity and Navier–Stokes equations. These equations represent the laws of conservation of different physical properties of a fluid in an Eulerian world. These equations follow some assumptions which include assuming the working fluid as a Newtonian and the flow as incompressible.

Continuity Equation:

$$\frac{\partial}{\partial x_i}(\rho u_i) = S_m = 0 \qquad (10.4)$$

3-D momentum equations for steady incompressible flow:

$$\rho u_j \frac{\partial}{\partial x_j}(u_i) = -\frac{\partial p}{\partial x_i} + \frac{\partial}{\partial x_j}\left[\mu \left(\frac{\partial}{\partial x_j}(u_i) + \frac{\partial}{\partial x_i}(u_j) \right) \right.$$
$$\left. -\frac{2}{3}\mu\delta_{ij}\frac{\partial}{\partial x_i}(u_i) \right] + \frac{\partial}{\partial x_j}\left(-\overline{\rho u_i u_j} \right) \tag{10.5}$$

Equation (10.5) is Navier–Stokes equations with supplementary Reynolds stress term. This Reynolds stress term is further modeled using the Boussinesq hypothesis which is present in Equation (10.6):

$$-\overline{\rho u_i u_j} = \mu_t \left(\frac{\partial}{\partial x_j}(u_i) + \frac{\partial}{\partial x_i}(u_j) \right) - \frac{2}{3}\delta_{ij}\left(\mu_t \frac{\partial}{\partial x_i}(u_i) + \rho k \right) \tag{10.6}$$

where μ_t is the turbulent eddy viscosity, k is the turbulent kinetic energy, and δ_{ij} is the Kronecker delta.

10.5 RESULTS AND DISCUSSION

From the qualitative analysis of pressure and velocity contours, we observe that there is a region of high pressure at the leading edge (stagnation point) and a region of low pressure on the lower surface of the rear wing assembly. This is found to be true for all the cases analyzed during the present study (AOA –14° to 16° and $v = 22$ m/s, 43 m/s, and 104 m/s). However, to avoid repeatability, contours for –2° of angle of attack (AOA) at 43 m/s have been presented (Figures 10.3–10.4).

From the Bernoulli equation, we know that whenever there is a high velocity, we have low pressure and vice versa. The above contours support this outcome. The pressure on the lower surface of the airfoil is found to be lower than that of the incoming flow stream and as a result, it effectively generated lift force in the downward direction, normal to the incoming flow stream.

Additionally, exhaustive quantitative analysis has been performed to identify the most effective AOA for maximum downforce keep the drag force to the minimum. Table 10.1 represents the variation of the ratio of lift and drag coefficients with AOA at all the three velocities analyzed during the present study.

The variations of lift and drag coefficients along with their ratio with AOA are also studied graphically during this study. However, again to avoid repeatability variation at 43 m/s only is represented in Figures 10.5–10.7.

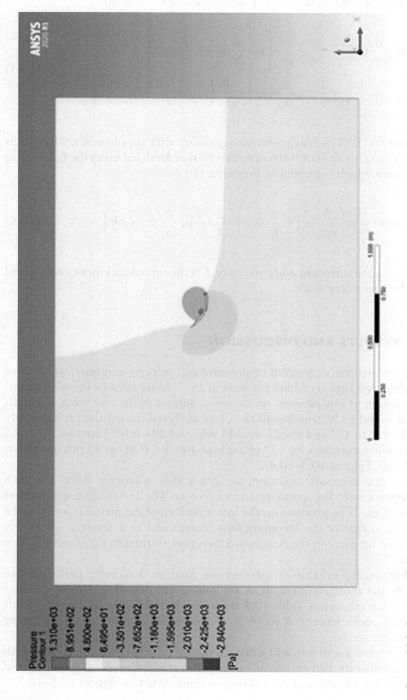

Figure 10.3 Static pressure contour for −2° AOA at 43 m/s.

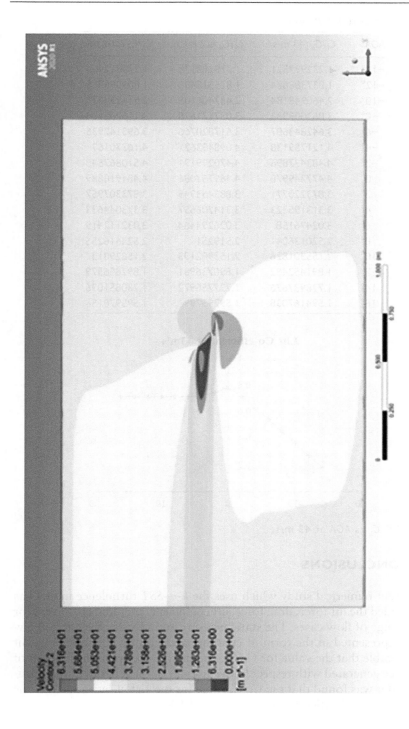

Figure 10.4 Velocity contour for −2° AOA at 43 m/s.

Table 10.1 C_L/C_d vs AOA

AOA	C_L/C_d (43 m/s)	C_L/C_d (22 m/s)	C_L/C_d (104 m/s)
−14°	1.232997521	1.218288178	1.270812464
−12°	1.837388624	1.815315488	1.889095335
−10°	2.460948184	2.437406105	2.516230773
−8°	3.063009933	3.037171628	3.116002104
−6°	3.642843687	3.617820766	3.693147935
−4°	4.121759138	4.098493621	4.16370169
−2°	4.483437896	4.470299191	4.510867548
0°	4.477349978	4.469757086	4.484918889
2°	3.873220771	3.883453945	3.873307957
4°	3.313195323	3.314706657	3.323044633
6°	3.02476158	3.026294464	3.032112419
8°	2.520136041	2.5192510	2.524616255
10°	2.155301826	2.153982193	2.15829813
12°	1.891452482	1.890736981	1.892856579
14°	1.728937676	1.727756972	1.730851096
16°	1.594167038	1.59295799	1.595978156

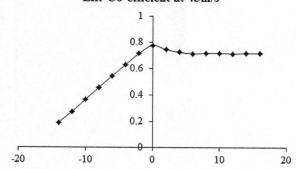

Lift Co-efficient at 43m/s

Figure 10.5 C_L vs AOA at 43 m/s.

10.6 CONCLUSIONS

The current numerical study which uses the *k-ω-SST* turbulence model has apprehended the lift force, drag force, lift coefficient, and drag coefficient for a wide range of flow cases. The static pressure and velocity for selected few cases are presented in the form of contours and graphs. It is quite evident from the table that the value for C_L/C_d, i.e. the ratio between the amount of downforce generated with respect to the amount of lift generated was calculated, and it was found that the C_L/C_d value for the airfoil set at −2° was the highest for the velocities 43 m/s, 22 m/s, and 104 m/s.

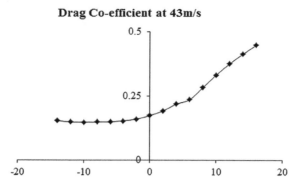

Figure 10.6 C_d *vs AOA at 43 m/s.*

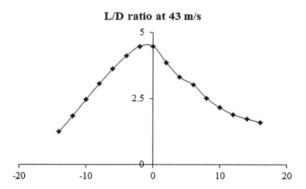

Figure 10.7 L/D vs AOA at 43 m/s.

Therefore, it is concluded that –2° is the optimum angle of attack as it generates more downforce keeping the drag force to the minimum.

REFERENCES

1. Triya Nanalal Vadgama, Arpit Patel, Dipali Thakkar, and Jignesh Vala, "Computational fluid flow analysis of formula 1 racing car." *International Journal for Scientific Research & Development*, 3(2), 2015: 1929–1936.
2. Parvathy Rajendran and Merryisha Sweety Gracia Samuel, "CFD validation of NACA 2412 airfoil." *IEEE* 2019. DOI: 10.13140/RG.2.2.16245.42723.
3. Meghana Athadkar and S. S. Desai, "Importance of the extent of far-field boundaries and the grid topology in the cfd simulation of external flows." *International Journal of Mechanical and Production Engineering*, 2(9), September 2014: 69–72.
4. A. R. S. Azmi, A. Sapit, A. N. Mohammed, M. A. Razali, A. Sadikin, and N. Nordin, "Study on airflow characteristics of rear wing of *F1* car." *IOP Conference Series: Materials Science and Engineering*, 243, 2017: 1–12.

Chapter 11

Numerical investigation on the melting and solidification of Bio-PCM having annulus eccentricity in horizontal double pipe thermal energy storage

Shubam Khajuria, Vikas and Ankit Yadav

Punjab Engineering College (Deemed to be University), Chandigarh, India

CONTENTS

11.1 INTRODUCTION

PCM stores a huge amount of thermal energy (melting) and releases (solidification) when required. The main problem with the use of PCMs (as storage medium) is their low thermal conductivity despite their high heat storage density at an almost constant temperature, chemical stability, and non-corrosive nature [1] Among different PCMs available, bio-PCM has various benefits such as high latent heat, highly abundant, low flammability, and no super cooling effect. PCMs have various applications in the restoration and usability of industrial waste energy, refrigeration and air conditioning systems, systems for solar energy storage, cooling and heating of buildings, electronic device's cooling, etc.

11.2 LITERATURE REVIEW AND OBJECTIVE

Various techniques on which researchers are concentrating their work to overcome this disadvantage include extended surfaces (fins) [2–3], the inclusion of nanoparticles (highly conductive materials) [4], metal foams [5–7], etc. Geometrical variations [8–9] (providing eccentricity) are preferred over other techniques as they do not reduce the PCM content and do not impact the economic activity for the model preparation. The dominance of natural convection for melting and conduction for solidification was illustrated by Ettouney et al. [10]. They experimented to investigate the characteristics of heat transfer in tube and shell heat exchangers having paraffin wax for both charging and discharging. They depicted that melting was natural convection dominated while solidification was conduction dominated.

Zhang and Faghri [11] studied numerically the effect on the discharging rate in a horizontal eccentric annulus. They concluded that the downward eccentricity decreases the rate of solidification. Yazichi et al. [9] found experimentally the outcomes of varying eccentricity on the performances of PCM in the double pipe heat exchangers. They concluded that varying the inner tube eccentricity downwards extends the melting natural convection zone. They also depicted that the increase in the eccentricity increases the solidification time.

According to the literature reviewed, the effect of eccentricity in the downward vertical direction for both melting and solidification is advantageous. With the downward eccentricity, the time for charging decreases and discharging increases. Barely any study was found in the literature that used Bio-PCM for LHESU. In this chapter, the various vertical eccentric positions of the inner heat transfer fluid tubes are studied with different magnitudes.

11.3 COMPUTATIONAL SECTION

11.3.1 Physical and numerical model

By providing the eccentricity to double pipe LHESU, different models are studied. A total of 10 cases are compared where cases 1, 2, 3, 4, and 5 depict concentric, 20 mm upwards, 40 mm upwards, 20 mm downwards, and 40 mm downwards, respectively, for melting and cases 6, 7, 8, 9, and 10 for solidification, respectively, with the same eccentricity as represented in Figure 11.1. The unstructured quadrilateral mesh was made by edge sizing, face meshing, and refinement techniques. Honeybee wax is used as a Bio-PCM for the study which is filled between the annulus.

Table 11.1 represents the thermo-physical properties required. To perform simulations ANSYS LUENT R19 software was used.

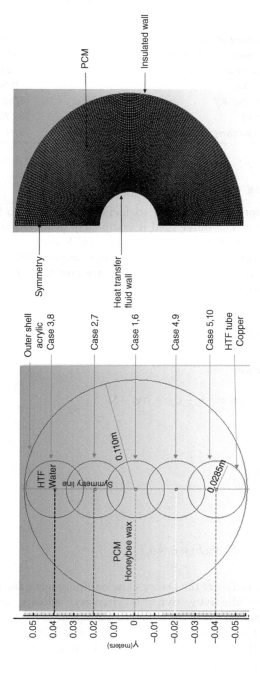

Figure 11.1 Representation of all the different geometrical arrangements of inner HTF tube with meshing of concentric case.

Table 11.1 Thermo-physical properties of honeybee wax [12, 13].

Property	Value
Dynamic viscosity (Pa-s)	0.02461
Thermal expansion coefficient β (1/K)	0.0001
Melting temperature range (°C)	62–64
Latent heat (J/kg)	247800
Specific heat (J/kg-K)	4760
Density (kg/m³)	789.75
Thermal conductivity (W/m-K)	0.41

11.3.2 Governing equations

Some of the assumptions considered to facilitate our thermo-fluid problem are the following: The outer shell is completely insulated and the heat energy loss is neglected, for the small changes in density, and Boussinesq approximation is considered for buoyant force impact. All the thermo-physical properties considered for our study are constant. The governing equations considered are:

Continuity:

$$\partial_t(\rho) + \partial_i(\rho\, u_i) = 0$$

Momentum equation:

$$\partial_t(\rho\, u_i) + \partial_j(\rho\, u_i u_j) = \mu \partial_{jj} u_i - \partial_i p + \rho g_i + S_i$$

where

$$S_i = C_{\text{mushy}}(1-\beta)^2 \frac{u_i}{\beta^3}$$

Energy equation:

$$\partial_t(\rho h) + \partial_t(\rho \Delta H) + \partial_i(\rho\, u_i h) = \partial_i(k \partial_i T)$$

Here, $S, \beta, k, \mu, u,$ and ρ are the Darcy's law damping term, the local liquid fraction, thermal conductivity, dynamic viscosity, velocity, and density, respectively. The S parameter is combined to the equation for momentum to add the effects of phase change during natural convection. C_{mushy} is mushy zone constant, and this constant usually varies between 10^4 and 10^7. In the present study, C_{mushy} is set to 10^5.

11.3.3 Boundary conditions and independence tests

The honeybee wax's initial temperature is assumed to be 333K and 338K for melting the solidification simulations, respectively. The heat transfer fluid temperatures are considered to be uniform at 353.15K and 313.15K for melting and solidification simulations, respectively. The heat transfer surface (HTS) is considered as a rigid boundary with a no-slip condition applied and is assigned a constant temperature boundary condition. For the time-step independence study, three time steps 0.01 s, 0.05 s, and 0.1 s are considered. From Figure 11.2(a) it can be clearly observed that there is only a small deviation in the melting fraction curve when there is an increase from 0.05 s to 0.1 s time step for the complete melting-solidification cycle. Therefore, for a stable solution of the computational model, 0.1 s time step is used.

The grid independence study includes three different numbers of grid elements 6000, 12000, and 24000. All the curves are very close to each other in Figure 11.2(b), so we concentrate our study on the complete melting-solidification cycle on 12,000 grid elements.

11.3.4 Model validation

For the validation, the numerical results of our study both in the case of melting and solidification are compared with the Mahdi et al. [14]. The dimensions considered are 70.7 mm for the outer shell's inner diameter and 50 mm for the inner HTF tube's outer diameter) and the model's conditions are taken similarly (Figure 11.3).

The PCM used for this validation is RT-50 (paraffin wax) and the present study results are matched with those of Mahdi et al. [14] from Figure. Thus, good reliability of results can be guaranteed by simulating the LTESU using the designed computational model.

11.4 RESULTS AND DISCUSSION

After completing the grid and time-independence tests along with the validation for charging as well as discharging, the results are shown in the form of contours and the transient variations in the parameters such as liquid fraction, enthalpy, and the average domain temperature are compared.

11.4.1 Melting and solidification

A series of computational simulations have been performed to compare the performance of different cases of LHESU.

Figure 11.4(a) shows the progress of melting in the form of contours temperature (left) and liquid fraction (right). Initially, it begins with the

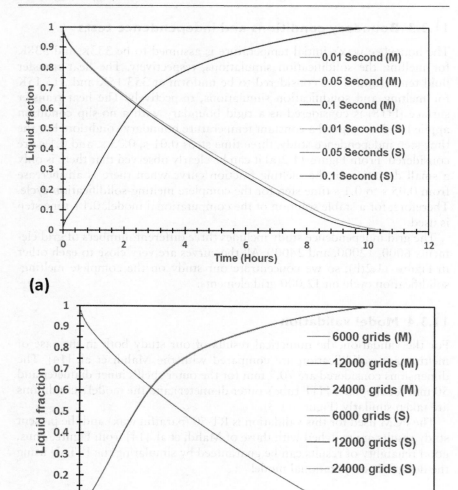

(a)

(b)

Figure 11.2 Independence test results of liquid fraction with variation of (a) time steps (b) number of grid elements.

conduction to the immediate PCM in contact. As PCM melts, natural convection starts playing a crucial role in further melting of PCM above the inner tube with the buoyancy effect. The lighter and warmer PCM moves upwards and the heavier colder PCM settles down. After the PCM in the

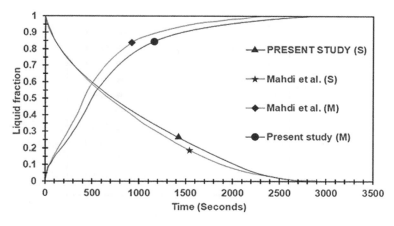

Figure 11.3 Comparison between the results of the present study and the results obtained by Mahdi et al. [14].

domain which is above the inner HTF tube gets melted, the natural convection's effect disappears and the thermal diffusion controls the left-over melting.

The graphical representation of every case 1–10 can be clearly depicted from Figure 11.4(a), (b), and (c) for liquid fraction, enthalpy, and average temperature, respectively. The energy stored and the average temperature values are considered only till the complete PCM either melts or solidifies. The 0 value of the liquid fraction represents the complete solid and 1 represents the complete liquid in Figure 11.4(a). The fastest and slowest melting processes were found in cases 5 and 3, respectively, and similarly, the fastest and slowest solidification processes were found in cases 6 and 10, respectively, shown in Figure 11.4(a).

The maximum average temperature was less for the fastest melting case compared to other cases for the melting process and it was the highest for case 4 until complete melting. For solidification, case 6 was found to be the best case as it requires the least solidification time, but the average temperature until complete solidification was the highest among the other cases, and it was the least for case 9 until complete solidification as shown in Figure 11.4(b). From Figure 11.4(c) it can be clearly interpreted that the enthalpy storage was the quickest for case 4, and the slowest for case 3 in case of melting. The discharge of energy was found to be the slowest and the quickest for cases 10 and case 6, respectively.

For the first half an hour, the melting is almost similar which is clear from the liquid fraction (right) and temperature (left) contours in Figure 11.5(a) for cases 1, 2, and 4. Case 4 has more PCM above the inner tube, so fast natural convection is more dominated compared to other cases. All the variations are compared with case 1, and it is decreased by 47.9% and 63.7%

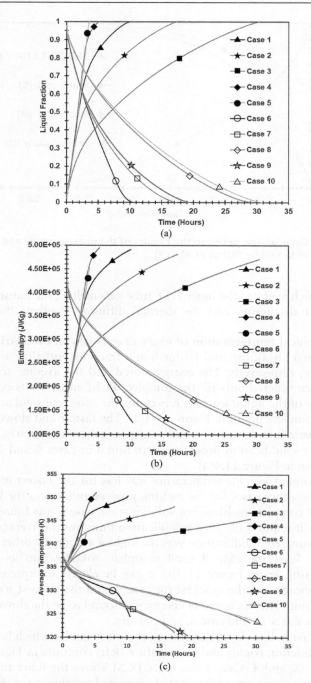

Figure 11.4 Representation of the variation of (a) liquid fraction, (b) enthalpy, and (c) average temperature with time.

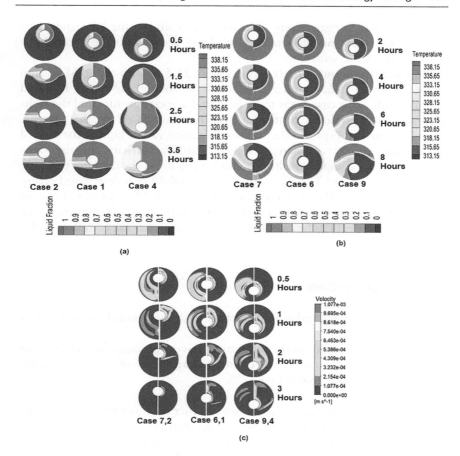

Figure 11.5 Representation of the contours for (a) melting (left-temp and right-
liquid fraction), (b) solidification (left-temp and right-liquid fraction),
and (c) velocity (left-solidification and right-melting).

for cases 4 and 5, respectively, and increased by 1.75 and 3.06 times for
cases 2 and 3, respectively. Figure 11.5(b) depicts the progress of the solidification for cases 6, 7, and 9.

For the solidification process, mainly conduction dominated and natural
convection has its impact when the solidification starts, but it's negligible
compared to conduction for the full discharging process. From the contours
in Figure 11.5(b), it can be clear that the progress of solidification front for
case 7 is in an upward direction and for case 9 it propagates in the downward direction but for concentric case, it moves symmetrically. So, the solidification time increases as the eccentricity increases as it is dominated by
conduction and the concentric case provides the best configuration to solid-
ify the PCM. Therefore, the solidification time increases by 1.85 and 2.95
times in cases 9 and 10, respectively. Similarly, 1.66 and 2.78 times increase

for cases 7 and 8, respectively. This increase is due to the lesser interface between the solid and molten PCM and the decrease in the area of the phase-change front with time.

The velocity contours are shown in Figure 11.5(c) with the solidification on the left for cases 6, 7, and 9 and melting on the right for cases 1, 2, and 4. The solidification case 6 has more balanced velocity profiles and the melting case 4 has the velocity profile for more time compared to cases 1 and 2. It can be clearly interpreted that the solidification rate is high in the case of upward eccentricity compared to downward eccentricity for the same amount. For the complete melting-solidification cycle, the overall cycle time increases by 1.20 and 1.69 times for 20 mm and 40 mm downward eccentricity, respectively. Also, it increases by 1.71 and 2.92 times in the case of 20 mm and 40 mm upward eccentricity, respectively. So, in our study, the concentric inner HTF tube case gives the best result for the complete cycle.

11.5 CONCLUSIONS

Some of the important and useful conclusions drawn from the present study are as follows:

The melting rate hikes as the inner tube moves vertically downwards due to the more and fast natural convection effect and vice versa. Case 5 reduces the most melting time that is by 63.7%.

- The rate of solidification increases whether the inner tube moves vertically upwards or downwards, but the increase is more in the case of downward eccentricity compared to upward eccentricity.
- For our study, the complete melting-solidification cycle performance is the best for case 1. Case 5 is the best for melting and case 1 for solidification.

A quick melting and slow-solidification thermal storage system is best suited for use in the building sector so that there is a reduction in the air-conditioning energy requirement, but it is not ideal for electricity storage purposes. So, it depends on the requirement of the system over which one can choose the TES model.

ACKNOWLEDGEMENTS

The authors acknowledge the CAD lab in PEC (Chandigarh) for providing the software and the computational facilities required for the simulations.

NOMENCLATURE

PCM	Phase change material	–
T_S	Temperature of solid PCM	K
T_L	Temperature of liquid PCM	K
T_r	Reference temperature	K
T	Temperature	K
g	Acceleration due to gravity	m/s^2

REFERENCES

1. S. Seddegh, X. Wang, A. D. Henderson, Z. Xing, "Solar domestic hot water systems using latent heat energy storage medium: A review," *Renew. Sustain. Energy Reviews*, 49, pp. 517–533, 2015, doi: 10.1016/j.rser.2015.04.147.S.
2. O. K. Yagci, M. Avci, and O. Aydin, "Melting and solidification of PCM in a tube-in-shell unit: Effect of fin edge lengths' ratio," *J. Energy Storage*, 24, no. May, p. 100802, 2019, doi: 10.1016/j.est.2019.100802.
3. A. M. Abdulateef, S. Mat, K. Sopian, J. Abdulateef, and A. A. Gitan, "Experimental and computational study of melting phase-change material in a triplex tube heat exchanger with longitudinal/triangular fins," *Sol. Energy*, 155, pp. 142–153, 2017, doi: 10.1016/j.solener.2017.06.024.
4. M. Al-Jethelah, S. H. Tasnim, S. Mahmud, and A. Dutta, "Nano-PCM filled energy storage system for solar-thermal applications," *Renew. Energy*, 126, pp. 137–155, 2018, doi: 10.1016/j.renene.2018.02.119.
5. M. Bashar and K. Siddiqui, "Experimental investigation of transient melting and heat transfer behavior of nanoparticle-enriched PCM in a rectangular enclosure," *J. Energy Storage*, 18, no. March, pp. 485–497, 2018, doi: 10.1016/j.est.2018.06.006.
6. H. I. Mohammed, P. T. Sardari, and D. Giddings, "Multiphase flow and boiling heat transfer modelling of nanofluids in horizontal tubes embedded in a metal foam," *Int. J. Therm. Sci.*, 146, no. August, p. 106099, 2019, doi: 10.1016/j.ijthermalsci.2019.106099.
7. C. Y. Zhao, W. Lu, and S. A. Tassou, "Flow boiling heat transfer in horizontal metal-foam tubes," *J. Heat Transfer*, 131, no. 12, pp. 1–8, 2009, doi: 10.1115/1.3216036.
8. M. J. Hosseini, M. Rahimi, and R. Bahrampoury, "Experimental and computational evolution of a shell and tube heat exchanger as a PCM thermal storage system," *Int. Commun. Heat Mass Transf.*, 50, pp. 128–136, 2014, doi: 10.1016/j.icheatmasstransfer.2013.11.008.
9. M. Y. Yazici, M. Avci, O. Aydin, and M. Akgun, "On the effect of eccentricity of a horizontal tube-in-shell storage unit on solidification of a PCM," *Appl. Therm. Eng.*, 64, no. 1–2, pp. 1–9, 2014, doi: 10.1016/j.applthermaleng.2013.12.005.

10. H. M. Ettouney, I. Alatiqi, M. Al-Sahali, and S. A. Al-Ali, "Heat transfer enhancement by metal screens and metal spheres in phase change energy storage systems," *Renew. Energy*, 29, no. 6, pp. 841–860, 2004, doi: 10.1016/j. renene.2003.11.003.

11. Y. Zhang and A. Faghri, "Analysis of freezing in an eccentric annulus," *J. Sol. Energy Eng. Trans. ASME*, 119, no. 3, pp. 237–241, 1997, doi: 10.1115/1.2888025.

12. H. Umar, S. Rizal, M. Riza, and T. M. I. Mahlia, "Mechanical properties of concrete containing beeswax/dammar gum as phase change material for thermal energy storage," *AIMS Energy*, 6, no. 3, pp. 521–529, 2018, doi: 10.3934/ ENERGY.2018.3.521.

13. N. Putra, E. Prawiro, and M. Amin, "Thermal properties of beeswax/CuO nano phase-change material used for thermal energy storage," *Int. J. Technol.*, 7, no. 2, pp. 244–253, 2016, doi: 10.14716/ijtech.v7i2.2976.

14. M. S. Mahdi, H. B. Mahood, A. F. Hasan, A. A. Khadom, and A. N. Campbell, "Numerical study on the effect of the location of the phase change material in a concentric double pipe latent heat thermal energy storage unit," *Therm. Sci. Eng. Prog.*, 11, no. March 2019, pp. 40–49, 2019, doi: 10.1016/j. tsep.2019.03.007.

15. K. Bhagat, M. Prabhakar, and S. K. Saha, "Estimation of thermal performance and design optimization of finned multitube latent heat thermal energy storage," *J. Energy Storage*, 19, no. January, pp. 135–144, 2018, doi: 10.1016/j. est.2018.06.014.

16. M. Gorzin, M. J. Hosseini, M. Rahimi, and R. Bahrampoury, "Nano-enhancement of phase change material in a shell and multi-PCM-tube heat exchanger," *J. Energy Storage*, 22, no. December 2018, pp. 88–97, 2019, doi: 10.1016/j.est.2018.12.023.

17. W. W. Wang, K. Zhang, L. B. Wang, and Y. L. He, "Numerical study of the heat charging and discharging characteristics of a shell-and-tube phase change heat storage unit," *Appl. Therm. Eng.*, 58, no. 1–2, pp. 542–553, 2013, doi: 10.1016/j.applthermaleng.2013.04.063.

18. M. Avci and M. Y. Yazici, "Experimental study of thermal energy storage characteristics of a paraffin in a horizontal tube-in-shell storage unit," *Energy Convers. Manag.*, 73, pp. 271–277, 2013, doi: 10.1016/j.enconman.2013.04.030.

19. Z. J. Zheng, Y. Xu, and M. J. Li, "Eccentricity optimization of a horizontal shell-and-tube latent-heat thermal energy storage unit based on melting and melting-solidifying performance," *Applied Energy*, 220, no. March, pp. 447–454, 2018, doi: 10.1016/j.apenergy.2018.03.126.

Chapter 12

Multi-parametric optimization of a convective type waste heat recovery system

Abhishek Arora and Rajesh Kumar

National Institute of Technology Kurukshetra, Kurukshetra, India

CONTENTS

12.1 INTRODUCTION

With the huge industrial growth and development that the world today is witnessing, the steel consumption of the world has risen many folds. India being the second largest producer of steel in the world is also following the same trend. Owing to these increased demands, the organizations all around the world have increased their steel production capacities and are trying to develop new methods to further improve their production outputs. Waste heat recovery is an important parameter considered for streamlining production and reducing wastage in steel-producing industries. While the performance of a heat recovery equipment in an industry depends on several factors such as outlet temperature of flue gases, flow rate of combustion air, etc., it is practically not possible to control all the factors.

12.1.1 Estimation of heat recovery

The purpose of a waste heat recovery process is to apprehend and transport waste heat from a fluid or solid and send it back into the combustion process as extra energy. The residual sources of heat mainly involve losses of heat by either of the processes of conduction, convection, radiation, or a combination of any of these. There are three main categories in which waste heat in industries is classified:

1. Waste heat at high temperature (at 400°C or above).
2. Waste heat at medium temperature (between 100°C and 400°C).
3. Waste heat at low temperature (below 100°C).

Heat at high temperature mostly comes directly from the combustion process; waste heat at medium temperature comes from the exhaust of the combustion process; and waste heat at low temperature comes from the heat treatment units or processes. There are different standard parameters being set to gauge the extent and grade of waste heat recovered, namely, quantity, quality, and temporal availability. The quantity of waste heat can be estimated using the below relation:

$$Q = \dot{V} \times \rho \times C_p \times \Delta T \tag{12.1}$$

where
Q = available heat content (Watt)
V = volumetric flow rate of the material transporting heat (m³/s)
ρ = material density (kg/m³)
C_p = material specific heat (J/kg-K)
ΔT = temperature difference responsible for heat transfer (K).

According to the second law of thermodynamics, the usability of energy depends on the condition of energy. The same amount of energy at high temperature holds more significance than that at a lower temperature. Hence, the quality of the heat recovered depends on the temperature at which it is available. Recovery of waste heat from low temperature sources is not only strenuous but also involves use of additional equipment such as pumps, blowers, etc., to improve the temperature condition and for making it suitable for heat recovery. The final parameter, i.e., temporal availability, quantifies the energy based on its accessibility at the time of its need. It is an important examination as it tries to match the availability of waste heat with the total load of the system, thereby influencing the effectiveness of the system. The waste heat recovery system considered for this study is a convective type recuperator. A typical convective recuperator involves the flow of combustion air within a set of parallel tubes arranged either in an inline or staggered manner and the flue gases flowing outside the tubes in a cross-flow pattern.

12.2 OBJECTIVES OF THE PRESENTED RESEARCH WORK

The presented research work focuses on the following objectives:

- Performing an optimization study for the convective recuperator currently installed at Continuous Strip Production Mill at ArcelorMittal/

Nippon Steel India Ltd., Hazira, Surat, India, suggests parametric and geometric modifications in the recuperator resulting in performance improvement of the device and waste heat losses reduction.

- To study the effects of variation of combustion air mass flow rate, number of tubes in the recuperator and tube external diameter on overall pressure drop in the device and convective heat transfer coefficient.
- To determine the optimum tube diameter offering a balance between Logarithmic Mean Temperature Difference (LMTD) and convective heat transfer coefficient as both parameters directly affect the heat transfer rates.
- To perform a comparative analysis between inline and staggered tube configuration and determine which configuration will be best suited for the device under consideration.
- Use of Computational Fluid Dynamics to account the effect of tube geometry on waste heat recovery and flue gas retention time and determine the recuperator configuration offering maximum waste heat recovery and effectiveness with the constraints of flow and geometric parameters as provided by the organization.

12.3 METHODOLOGY

We defined our objectives in-line with the optimization of the recuperator that could improve its current performance or at least suggest alterations in its current operational and geometric parameters that may be incorporated in future for its performance improvement. Since, cost of running a device plays a pivotal role in any industry, it is also essential to suggest methods that may prove to be cost effective in implementation. Kayabasi and Erdogan [1] designed a unique convective recuperator to increase the combustion air temperature entering in the hot stoves of the blast furnace. Simulations were carried out using Strelow's static simulation model which is based on ε-NTU method. To fulfill the objectives in hand, the entire research work for this study is carried out using two different methodologies. A detailed description of both methodologies is discussed in the following sections.

12.3.1 Theoretical study and analysis

The theoretical study and analysis of the recuperator will attempt to achieve the following objectives:

1. To study the effect of number of tubes in the recuperator, combustion air mass flow rate, and tube external diameter on overall pressure drop in the recuperator and convective heat transfer coefficient.
2. To determine an optimum tube external diameter balancing the convective heat transfer coefficient and LMTD.

3. To do a comparative analysis between inline and staggered tube arrangement and determine which arrangement will be best suited for the device and conditions under consideration.

The entire recuperator is divided into two different sections: 1. Cold section – the section of recuperator in which combustion air at atmospheric temperature and pressure is inducted in the recuperator and after flowing through the diffuser and the tube section reaches the recuperator bottom. 2. Hot section – section in which combustion air from the recuperator bottom flows in the other tube section and diffuser and from there is sent to the furnace section to carry out the reduction reactions (Figure 12.1). The air while flowing within the tubes will get preheated by the flue gas flowing

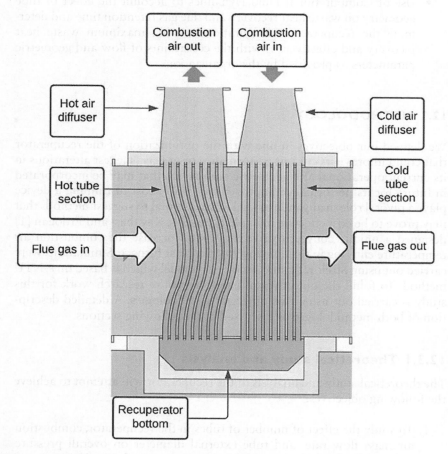

Figure 12.1 Recuperator model indication flow direction of the fluids.

outside the tubes within the recuperator shell in a cross-flow direction. The schematic shown in Figure 12.1 describes the flow of both fluids. The upward movement of the combustion air occurs due to density changes as it gets heated during the flow.

The recuperator in the actual working condition has 440 tubes in total, 220 each side divided as 20 rows along the transverse direction and 11 tubes per row along the longitudinal direction. For calculation purpose, the above-mentioned two sections are further divided into two sections each, namely, cold air diffuser – diffuser in the cold section, cold tubes- tubes in the cold section, hot tubes – tubes in the hot section and hot air diffuser – diffuser in the hot section. At the entry and exit of all the sections mentioned above, pressure, velocity, and friction head loss calculations are done by using the equations such as Continuity equation, Colebrook equation, Bernoulli's equation, and Zhukauskas relation. Ludwig and Zajac [2] also carried out a recuperator modification using uniformity of distribution of flue gas in the space between the tubes and pressure drop as the criteria for modification. The final results led to an improvement in gas flow parameters due to a more homogeneous inter-pipe gas flow and smaller pressure drop. It is pertinent to mention that the flow of combustion air in the tubes is taken as a steady flow and tubes in both the sections are considered as straight while making the calculations.

The experiment was performed on the recuperator model using the operational and geometric data provided by the organization. The flue gas composition shown in Table 12.1 was used to determine the thermophysical properties of the flue gas. The thermophysical properties and the experimental data used are shown in Tables 12.2 and 12.3 for reference. The results obtained from the experiment were further utilized for the validation of the theoretical results obtained from the current study. To undertake the theoretical study, an analytical tool was developed in MS-Excel to perform the calculations for different values of number of tubes in the recuperator, combustion air mass flow rate and tube external diameter. It is to be noted that while varying any of the above three parameters, the remaining two parameters were treated with a fixed value for the calculation. For example, when the calculations were done for different number of tubes in the recuperator, the values of mass flow rate and tube external diameter were kept fixed for this set of calculations. Same procedure was followed for the remaining two parameters as well. To account for the head loss due to friction, the friction factor was calculated using the Colebrook's equation iteratively until a

Table 12.1 Flue gas composition (data taken from experimental analysis)

Symbol	N_2	O_2	H_2	CO	CO_2
% (volumetric)	66.8	1.6	0.8	1.2	29.6

Table 12.2 Thermophysical properties of air and flue gas

Properties of air (at 30°C)		Properties of flue gas (at 762.75°C)	
Inlet temperature	30°C	Inlet temperature	762.75°C
Outlet temperature	524.9°C	Outlet temperature	622.70°C
Density (ρ)	1.225 kg/m³	Density (ρ)	0.504 kg/m³
Viscosity (μ)	1.81E-05 Pa-s	Viscosity (μ)	3.74E-05 Pa-s
Specific heat (Cp)	1.0049 KJ/kg-K	Specific heat (Cp)	1.150 KJ/kg-K
Prandlt number (Pr)	0.707	Prandlt number (Pr)	0.768
Thermal conductivity (k)	0.024 W/m-K	Thermal conductivity (k)	0.056 W/m-K
Mass flow rate (m)	0.703 kg/s	Mass flow rate (m)	2.31 kg/s

Table 12.3 Recuperator geometric parameters and thermal properties

Cold section		Hot section	
Diffuser area (top)	0.522 m²	Diffuser area (top)	0.817 m²
Diffuser area (bottom)	1.683 m²	Diffuser area (bottom)	1.683 m²
Tube grade	SS 409	Tube grade	SS 309
Density (ρ)	7610 kg/m³	Tube density (ρ)	8000 kg/m³
Thermal conductivity (k)	24 W/m-K	Tube thermal conductivity (k)	15.6 W/m-K
Specific heat (Cp)	450 J/kg-K	Specific heat (Cp)	502.4 J/kg-K
Tube length (L)	2.25 m	Tube length (L)	2.25 m
Tube external diameter (d)	0.0445 m	Tube external diameter (d)	0.0445 m
Tube thickness (t)	0.003 m	Tube thickness (t)	0.003 m
Total no. of tubes	220	Total no. of tubes	220
Longitudinal pitch	0.075 m	Longitudinal pitch	0.075 m
Transverse pitch	0.1 m	Transverse pitch	0.1 m
Diffuser diameter	0.816 m	Diffuser diameter	1.02 m

residual of 10^{-4} is achieved after comparing the LHS and RHS of the equation.

To calculate the heat transfer coefficient on the flue gas side, the below relations were used from [3],

For $50 < \mathrm{Re}_D < 100$

$$Nu = 0.8 \times \mathrm{Re}_D^{0.4} \times \mathrm{Pr}^{0.36} \times (\mathrm{Pr}/\mathrm{Pr}_s)^{0.25} \qquad (12.2)$$

For $1000 < \mathrm{Re}_D < 2 \times 10^5$

$$Nu = 0.27 \times \mathrm{Re}_D^{0.63} \times \mathrm{Pr}^{0.36} \times (\mathrm{Pr}/\mathrm{Pr}_s)^{0.25} \qquad (12.3)$$

While the last term in both Equations (12.2) and (12.3) was dropped during the calculations.

To account for dynamic losses due to surface roughness of the tubes, a surface roughness value of 0.005 mm is considered for the tubes. Since, the current study does not focus on effect of material change of recuperator tubes, an estimated value of surface roughness as specified by the manufacturer is considered during the calculations of the friction factor.

The flow chart shown below in Figure 12.2 will help us in understanding how the calculations are performed with the help of analytical tool developed in MS-Excel.

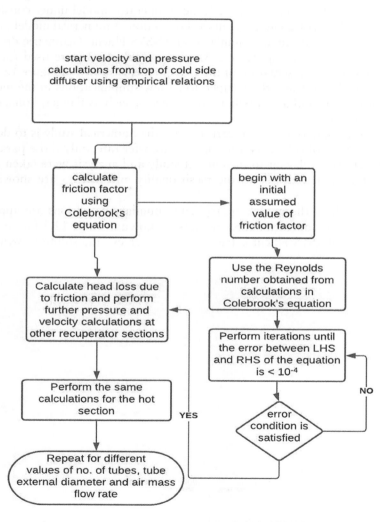

Figure 12.2 Flow-chart for the calculation of parameters in the analytical tool.

12.3.2 Numerical study and analysis

While the analytical model developed in MS-Excel was efficient in estimating the effects of variation in geometric and flow parameters on overall pressure drop of combustion air in the recuperator and convective heat transfer coefficient, the effect of tube shape on heat transfer rate could not be determined through it. Hence, to capture the effects of tube curvature or geometry on heat transfer rate, a numerical model of the recuperator was studied and analyzed. This may be attributed to the fact that a curved tube will allow the air to cover a larger distance as compared to a straight tube, thereby, allowing it to pick up more heat from the flue gas flowing outside the tubes. As shown in Figure 12.1, the recuperator model under consideration has tubes with a curvature in the hot section. A numerical model of the recuperator is modeled and simulated in ANSYS Fluent. Geometric dimensions of the actual recuperator as shown in Table 12.3 were used for the CAD model. A representation of the CAD model is shown in Figure 12.3.

Also, while modeling the recuperator, a few simplifications in the model were also made and any auxiliary attachments, such as flanges, bolts, etc., are not modeled.

Since our main objective of carrying out the numerical study is to determine the effect of tube curvature on heat transfer rate, only those parameters which were relevant to the current study and analysis were taken into consideration. The details of the mesh quality and metrics are shown in Table 12.4.

After meshing the model, appropriate boundary conditions are applied from the current operational parameters as shown in Table 12.2 for air and flue gas. An ANSYS Fluent solver is used to achieve the solution with the

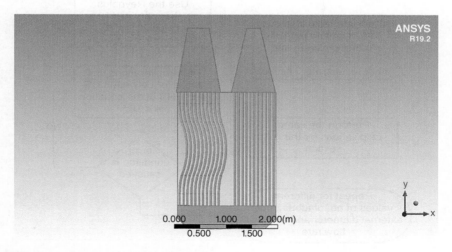

Figure 12.3 Recuperator CAD model.

Table 12.4 Mesh parameters

Mesh parameter	Value
No. of nodes	3132646
No. of elements	1.78E+07
Average skewness	0.25
Average element quality	0.81
Average aspect ratio	2.08
Average orthogonal quality	0.75

Table 12.5 Solver characteristics

Solver parameter	Value
Solver	Pressure based
Turbulence model	k-ω SST model
Boundary condition at inlet	Velocity based
Boundary condition at outlet	Pressure based
Boundary condition at walls	No slip, adiabatic; symmetry at front and back
Discretization	Second-order upwind

current model. The initial calculations of the flow parameters indicated the existence of the turbulent flow and hence k-ω SST turbulence model is used to obtain the solution.

The solver characteristics used for this study are presented in Table 12.5. The elementary stages of the study also included the optimization of the solver, and the values presented in Table 12.5 were attained after it.

Regarding the solution, a stable solution for the problem under consideration was attained after running the solution for 600 iterations in the solver and the calculations were continued until the values of continuity, u, v, w momentum, energy, turbulent kinetic energy, and turbulence dissipation rate became constant and the scaled residuals reached the values of 10^{-4}. The effectiveness calculation was a major aspect of the current study, and with numerical analysis, we were able to obtain the temperature values of flue gas and combustion air at the outlet while the values at the inlet were obtained as experimental data. The effectiveness calculations were done using the relations as provided below taken from Theodore et al. [4].

$$\varepsilon = \frac{\text{Actual heat transfer rate}}{\text{Maximum heat transfer rate}} \qquad (12.5)$$

Applying first law of thermodynamics to the heat exchanger,

$$\text{Actual heat transfer rate} = \dot{m}_h \times C_{Ph} \times (T_{hi} - T_{he}) = \dot{m}_c \times C_{Pc} \times (T_{ce} - T_{ci}) \quad (12.6)$$

$$\text{Maximum heat transfer rate} = (\dot{m} \times c)_{small} \times (T_{hi} - T_{ci}) \quad (12.7)$$

For the fluid properties undertaken for this study, $(\dot{m}_h \times C_{Ph}) > (\dot{m}_c \times C_{Pc})$ Hence, the effectiveness can be given as

$$\varepsilon = \frac{T_{ce} - T_{ci}}{T_{hi} - T_{ci}} \quad (12.8)$$

The same model equation was used to perform effectiveness calculations for all the three recuperator configurations, and the results were presented for a comparative analysis of the recuperator's performance. Effectiveness calculations were also performed using the results achieved by the experiment and it was used to validate the results of numerical analysis.

12.4 RESULTS AND DISCUSSIONS

We carried out both the studies in this research work to attain a certain set of predefined objectives which were important to the organization in which this research work was carried out. The theoretical study which involved the use of an analytical tool developed in MS-Excel helped us in understanding and evaluating the effects of variation of three parameters, namely, number of tubes in the recuperator, combustion air mass flow rate, and tube external diameter on the overall pressure drop of combustion air in the device and convective heat transfer coefficient. The results obtained from theoretical and numerical studies are presented in separate sections below while the results obtained from the experiment are used for the purpose of validation.

12.4.1 Results from the theoretical study

The first result obtained from the analytical tool shown in Figure 12.4 is achieved by varying the number of tubes in the recuperator. The total number of tubes in the recuperator is 440 divided as 220 each section. Since, for the analytical study, we have considered straight tubes on both the sections, we started our calculations with tube number of one section, i.e. 220, and subsequently varied the number of tubes to both upstream and downstream values to observe its effect on the overall pressure drop of combustion air in the recuperator. The results indicated an increase in the overall pressure drop with a decrease in the tube number. This can be attributed to the reason

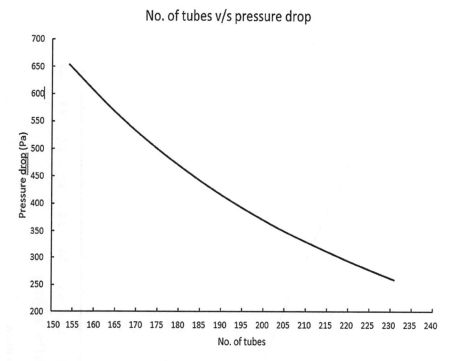

Figure 12.4 No. of tubes v/s pressure drop variation.

that while we are decreasing the number of tubes, the mass flow rate of air entering in the recuperator is taken at a fixed value.

The second result obtained by varying the combustion air mass flow rate and keeping the number of tubes and tube external diameter at a fixed value is shown in Figure 12.5. Calculations with an initial mass flow rate of 3.81 kg/s, which is the mass flow rate in the current operational device, were started and from there it was subsequently reduced by a factor of 2%–3%.

The result shown in Figure 12.6 is achieved by the variation of tube external diameter by keeping the total number of tubes and mass flow rate of combustion air at a fixed value. The effect of this variation is studied on the convective heat transfer coefficient which is an important factor in determining the overall heat transfer rate. For flow of a fluid over a bank of tubes, Re = 1000 is considered as the critical Reynolds number above which the fluid enters into a transition zone (laminar to turbulent) from the laminar zone. It is interesting to note that the tube external diameter in the experimental setup has a value of 44.5 mm and the Reynolds number corresponding to this diameter is in transition zone. A decrease in the tube external diameter indicated the switch of the flow toward the laminar zone, thereby reducing the heat transfer coefficient. The dynamic losses that may occur due to transition of flow away from the laminar zone were also considered.

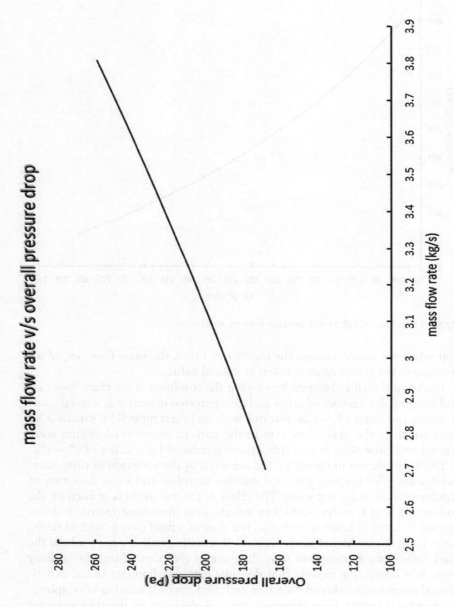

Figure 12.5 Combustion air mass flow rate v/s pressure drop variation.

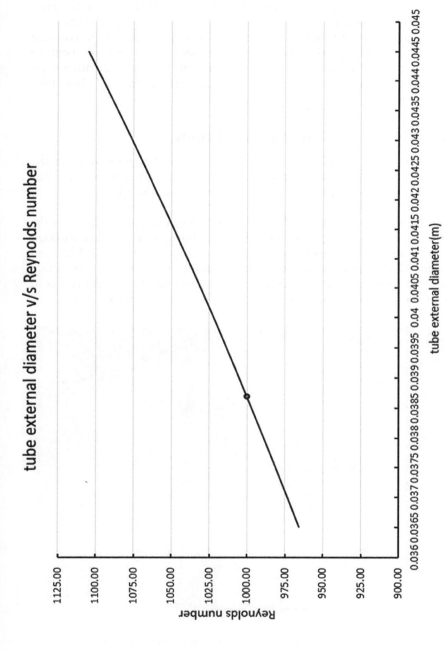

Figure 12.6 Tube external diameter v/s Flue gas Reynolds number.

Figure 12.6 indicates that the critical Reynolds number value of 1000 is achieved at a tube external diameter of 38.67 mm. Keeping the number of tubes to 220 on each section of the recuperator and reducing the tube external diameter resulted in a sudden reduction in heat transfer coefficient.

The results from the comparative analysis of an inline and staggered tube configuration indicate that higher heat transfer rates (improvement of 9.04% in convective heat transfer coefficient) can be achieved with a staggered tube arrangement instead of the currently followed inline tube arrangement.

12.4.2 Results from the numerical study

The results from the numerical study were used to determine whether the current recuperator configuration is optimal or an alternate recuperator configuration can offer better heat transfer rates and the effect of tube curvature on heat transfer.

Temperature contours can also give us a significant representation of the temperature variations of combustion air and flue gas as they flow through the recuperator. The temperature contours shown in Figures 12.7, 12.8, and 12.9 for all the three recuperator configurations undertaken for this study indicate a more uniform temperature distribution for the recuperator configuration with curved tubes in both the hot and cold sections.

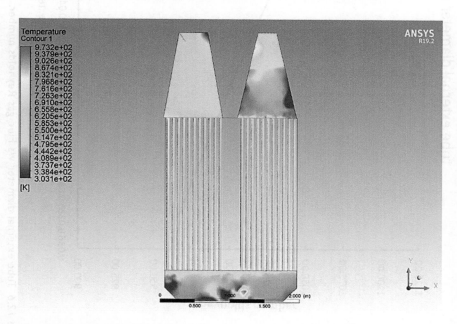

Figure 12.7 Temperature contours for straight tubes configuration.

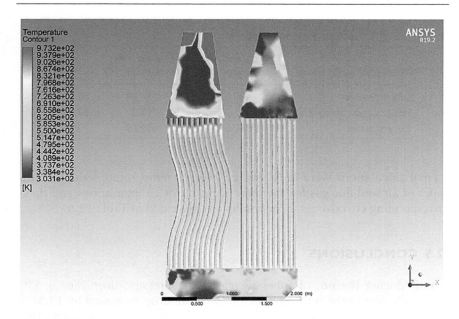

Figure 12.8 Temperature contours for straight + curved tubes configuration.

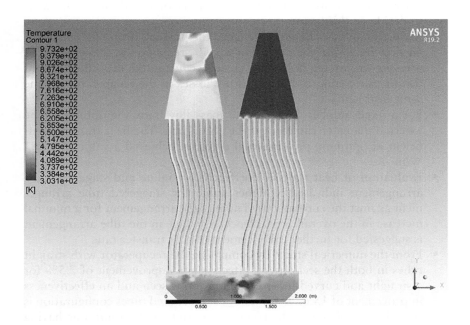

Figure 12.9 Temperature contours for curved tubes configuration.

Table 12.6 Outlet temperature and effectiveness values for different recuperator configurations

Recuperator configuration	Air outlet temperature (°C)	Flue gas outlet temperature (°C)	ε (%)
Straight tubes	488.92	756.84	62.63
Straight and curved tubes	523.43	709.3	67.34
Curved tubes	560.58	692.09	72.41

Final calculations were done for the average temperature values at the outlet of air and flue gas for the three recuperator configurations and their corresponding effectiveness. The results are presented in Table 12.6.

12.5 CONCLUSIONS

- Reducing the no. of tubes increased the pressure drop. For a 5% reduction in the tube number, the pressure drop increased by 13.5%. However, increasing the no. of tubes by 5% reduced the pressure drop by 12%, but it also led to an increase in the area by 4.56%. Hence, reducing the pipe number is not advisable; however, a cost analysis considering the area available and cost per tube will further allow us to decide on this factor.
- Reduction in mass flow rate reduced the pressure drop of combustion air flowing within the pipes. For a 2.5% reduction in mass flow rate, pressure drop reduces by 4%. Hence, reducing the mass flow rate of the air can compensate for pressure losses in the recuperator for the same heat load.
- For a fixed number of tubes and mass flow rate, the optimum pipe external diameter came out to be 38.67 mm. The heat transfer coefficient at optimum pipe external diameter achieved a value of 14.51 W/m²-K.
- Evaluation of heat transfer coefficient for inline and staggered tube arrangement indicated a higher value for staggered tube arrangement against the currently used inline tube arrangement for a minimal increase in the overall area. Hence, a change in the tube arrangement is suggested for further enhancement of heat transfer rate.
- From the numerical study, by comparing the recuperator with straight tubes in both the sections, an effectiveness improvement of 7.5% for a straight and curved tubes configuration is seen and an effectiveness improvement of 15.62% for a both sides curved tubes configuration is seen. Hence, changing the recuperator configuration into one having maximum heat recovery can save down fuel costs, thereby increasing the efficiency of the complete heat recovery system.

REFERENCES

1. Erhan Kayabasi and Selim Erdogan, "Design and simulation of recuperator for hot stoves", *International Journal of Mechanical and Production Engineering (IJMPE)*, 5(3), 2017, 15–20.
2. Wojciech Ludwig and Daniel Jajac, "Modification of a recuperator construction with CFD methods", *Chemical and Process Engineering*, 38(4), 2017, 567–576. doi: 10.1515/cpe-2017-0045.
3. M. Thirumaleshwar, *Fundamentals of Heat and Mass Transfer*, Pearson Education, 2006.
4. Theodore L. Bergman, Adrienne S. Lavine, Frank P. Incropera, and David P. Dewitt, *Fundamentals of Heat and Mass Transfer*, John Wiley and Sons Inc., 2011.

Chapter 13

Sump pump output flow parameter optimization by Taguchi method

Prakash Shinde and R.R. Arakerimath

G.H. Raisoni College of Engineering and Management, Pune, India

CONTENTS

13.1 INTRODUCTION

The pump sump studies are mainly conducted for the optimization of the geometry and pump suction hydraulic parameters. The proper hydraulic design of the sump is important for the smooth flow entry and proper suction parameters in the pump inlet. This original and novel work is carried out considering the real-time problems at the field, and various simulations are conducted to understand the dominant parameters in the pump suction hydraulic design. It is always a difficult task to design and predict the pump sump hydraulic geometry as large number of parameters are involved in the same. With respect to this predicting the performance of pump sump with trial and error method is a time-consuming task and has huge cost impact on the project as pump sump is mostly in the civil work. Apart from this, CFD analysis is widely used in the sump design and optimization [1–3]. Pump sump is chosen in this analysis because it is mostly used in all water supply schemes, etc. For performance improvement and minimizing the losses such as turbulence, vortices, and recirculation in the flow at entry in the pump many studies are conducted [4, 5, 11]. Taguchi method is a robust method for the optimization and systematic approach with design of the experiments involving the various input parameters to achieve the required quality parameters [6, 7]. Taguchi approach is easy for implementation. Taguchi method is sensitive to input parameters and creates the robust designs with optimum time and minimal cost [8–10, 12]. In sump optimization process, pump inlet velocity, well inlet velocity, etc. are attained by the various parameters. These parameters are used for each optimization process based on the pump center-to-center distance, no. of pumps installed in the sump.

Pump sump optimization parameters are as follows:

a. Inlet pipe velocity – This is the velocity of the approaching pipe/channel in the pump sump. This is governed by the pump flow rate and pipe diameters.
b. Inlet Bell velocity – This is the velocity of the pump inlet governed by the bell mouth diameter of the pump.
c. Pump center-to-center distance – Pump center-to-center distance is governed by the installation arrangement in the pump sump and bottom bell clearance with respect to adjacent bell clearance.
d. No. of Pumps – No. of pumps installed in well is to be considered along with the side wall clearance of the pump on the extreme sides in the pump sump.

13.2 LITERATURE REVIEW AND OBJECTIVE

In the era of technological advancement, optimization plays important role in the pump sump hydraulic design and its effect on the pump performance. Various studies were conducted to ascertain the better hydraulic performance of the pump sump. Such studies included the computational fluid analysis and physical sump model studies for the varied pump sump geometries. Many studies were concerned about the various parameters affecting the pump optimization for the flow smoothing [1–4]. The various parameters affecting the pump performance are from the suction side hydraulic conditions. These parameters are related to pump installation dimensions in the sump and other side clearances. The objective of this study is to optimize the pump suction side hydraulic performance by optimizing the sump geometry and ascertain the significance level of the independent parameters. The main challenge during the optimization is the turbulence in the flow field. In the present study, four parameters are identified for the effect on the pump

Table 13.1 Orthogonal array design

	Factor	Min	Normal	Max	Unit
Factor 1	Well inlet velocity	0.5	1.2	2	m/s
Factor 2	Bell inlet velocity	0.5	1.5	2.5	m/s
Factor 3	Pump c/c distance	1D	2D	3D	m
Factor 4	No. of pumps	2	3	5	Nos.

Table 13.2 Orthogonal array design

Well inlet velocity, m/s	Bell inlet velocity, m/s	Pump c/c distance, m	No. of pumps, Nos.
1	1	1	1
1	2	2	2
1	3	3	3
2	1	2	3
2	2	3	1
2	3	1	2
3	1	3	2
3	2	1	3
3	3	2	1

sump with the help of the various literature survey conducted for the pump sump studies and optimization (Table 13.1).

13.3 TAGUCHI METHOD FOR DESIGN OPTIMIZATION

3D modeling is done using the Creo Parametric 3D modeling tool for the pump sump geometries. Various geometries are created as mentioned in the design of experiment (DOE) as shown in Table 13.2.

13.4 FACTOR AND LEVELS FOR TAGUCHI METHOD

As discussed earlier, the below-mentioned four factors – well inlet velocity, bell inlet velocity, pump c/c distance, and no. of pumps – are selected with the three levels, i.e. minimum, normal, and maximum as mentioned in Table 13.1 as per the previous literature data.

Along with the above-mentioned parameters, the optimal range is selected for the levels as mentioned above. The proper orthogonal matrix is decided from the above-mentioned parameters. The levels considered in this method were three levels, i.e. minimum, nominal and maximum. The suitable L9 (3^4)

orthogonal arrays, four factor and three levels were selected for analyzing the Taguchi method. The range of the levels selected optimally for the above, i.e. minimum, nominal and maximum values as mentioned in the Table 13.1 for forming the Taguchi orthogonal array.

13.5 DESIGN OF EXPERIMENT USING ORTHOGONAL ARRAY

Design of the orthogonal array having the four factor and three levels and having the 9 number of experiment are considered. This Taguchi L9 DOE method gives 99.96% accurate results as per literature survey, hence is more reliable [6, 7]. L9 Orthogonal DOE is used to estimate the main effects that are independent of two-factor interactions and give full precision on the estimation of such effects since these effects are not correlated to each other. Full factorial design also can be used for further DOE as experiments for more study.

13.6 DESIGN OF EXPERIMENT AS PER ORTHOGONAL L9 TABLE

3D model is created for the sump geometry as per the below parameters. The parameters are listed as well inlet velocity, bell inlet velocity, pump installation distance from the adjacent pumps, and no. of the pumps installed in the sump. These parameters are having the major influence on the hydraulic condition of the pump flow entry in the pump bell mouth. These parameters, if not controlled properly, may have effect on the pump hydraulic performance (Table 13.3).

Table 13.3 Design of experiment – L9 Table

Well inlet velocity, m/s	Bell inlet velocity, m/s	Pump c/c distance, m	No. of pumps, Nos.
0.5	0.5	1	2
0.5	1.5	2	3
0.5	2.5	3	5
1.2	0.5	2	5
1.2	1.5	3	2
1.2	2.5	1	3
2	0.5	3	3
2	1.5	1	5
2	2.5	2	2

13.7 SIMULATIONS PERFORMED

As mentioned in Table 13.3, various experimental simulations were performed with 3D model in Creo Parametric. These models analyzed using the Ansys Fluent software and computational plots are taken to form the response variables as pump output flow and tangential velocity components in the pump inlet bell area.

13.8 MODELING AND ANALYSIS OF THE PUMP SUMP

As shown below 3D Model is created for the DOE table No. 13.3 along with the parameters mentioned. In model creation, hydraulic dimensions are maintained as per the actual installation at side and sump parameters affecting the pump performance are maintained.

In the above mentioned Figure 13.1, two nos. of the pumps are mounted in the well and suction inlet of the rectangular size as per the mentioned

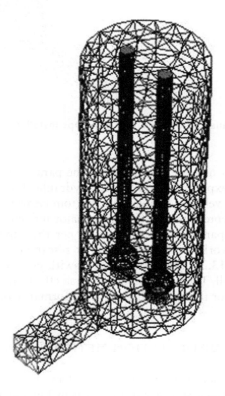

Figure 13.1 3D modeling and meshing of the 2 nos. of pumps installed in circular well.

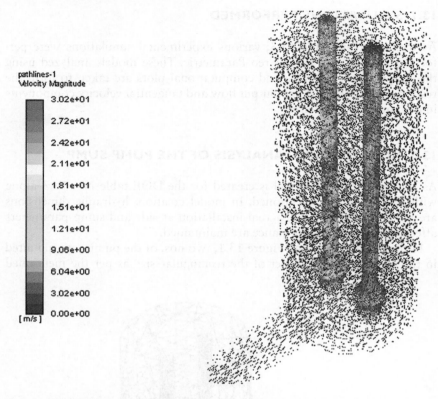

pathlines-1
Velocity Magnitude

3.02e+01	
2.72e+01	
2.42e+01	
2.11e+01	
1.81e+01	
1.51e+01	
1.21e+01	
9.06e+00	
6.04e+00	
3.02e+00	
[m/s] 0.00e+00	

Figure 13.2 CFD analysis of the 2 nos. of pumps installed in circular well.

design of experiment table is created. All the parameters are considered as per the design of experiment table for the hydraulic design of the sump. The inlet suction pipe velocity, bell velocity, bottom clearances, side clearances, pump center-to-center distance is having major influence on the pump performance. These parameters are varied as per the experimental data and earlier research work carried out by other researchers. The effects are as shown in Figure 13.2, vortex, turbulence, swirl, peroration in the entry of the pump inlet bell. These mentioned adverse effects are presents and have to be minimized for the proper hydraulic performance of the pump.

13.9 TAGUCHI OPTIMIZATION METHOD

The results calculated from the simulation data are mentioned herewith as pump flow output as response parameters as in Taguchi design (Table 13.4).

Table 13.4 Factor and response table

F1	F2	F3	F4	Pump flow, m3/h
0.5	0.5	1	2	2345
0.5	1.5	2	3	2510
0.5	2.5	3	5	2490
1.2	0.5	2	5	2430
1.2	1.5	3	2	2540
1.2	2.5	1	3	2380
2	0.5	3	3	2530
2	1.5	1	5	2330
2	2.5	2	2	2450

13.10 RESULTS AND DISCUSSION

The results are discussed in details for both the response parameters and factor in detail.

13.10.1 Data analysis for the response table of pump output flow

Table 13.5 mentions the responses as per the Minitab statistical analysis. The response variable pump output flow is larger the better as it has to be maximized as per the various input factor level variation (Tables 13.5 and 13.6).

Figure 13.3 shows the main effect plot. As shown in the below chart, the main inlet velocity increases from 1 m/s to 1.2 m/s and then it decreases further. The bell inlet velocity of 1.5 m/s is centric, whereas the pump c/c distance of 3D is more favorable. The no. of the pump flow increases from 2 Nos. to 3 Nos. and further reduces the pump out flow performance.

Table 13.5 Response table for signal-to-noise ratios

Level	Well inlet velocity, m/s	Bell inlet velocity, m/s	Pump c/c distance, m	No. of pumps, Nos.
1	67.77	67.73	67.43	67.76
2	67.78	67.81	67.83	67.86
3	67.73	67.75	68.03	67.66
Delta	0.05	0.09	0.60	0.20
Rank	4	3	1	2

Table 13.6 Response table for means

Level	Well inlet velocity, m/s	Bell inlet velocity, m/s	Pump c/c distance, m	No. of pumps, Nos.
1	2448	2435	2352	2445
2	2450	2460	2463	2473
3	2437	2440	2520	2417
Delta	13	25	168	57
Rank	4	3	1	2

The regression equation for the output flow response is as below.

13.10.2 Regression equation

Pump flow = 2323.2 – 8.0 Well inlet velocity, m/s + 2.5 Bell Inlet Velocity, m/s + 84.2 Pump c/c distance, m – 12.14 No. of Pumps, Nos.

13.10.3 Model summary

S	R-sq	R-sq(adj)	R-sq(pred)
36.6687	89.29%	78.57%	13.25%

From the ANOVA Table 13.8, it is observed that Factor 3 has significant effects (at 95% Confidence Level, CL) on both the mean and variation of the response variable pump output flow.

The maximum effect of the pump c/c distance parameters is having about, 42504.2/50200 = 84.66% and the second most significant parameter is no. of pump installed, i.e. 2064.3/50200 = 4.11% as from the above data. Other parameters are not that much significant in the pump output flow response.

13.11 CONCLUSIONS

Parameters are analyzed and optimized as per Taguchi orthogonal experimental method. The pump sump optimization results are obtained by analysis as shown in Tables 13.7 and 13.8. Four numbers of hydraulic parameters are considered with the three levels of variation in this optimization study. As per the above analysis, pump center-to-center distance is the highest influence on the pump flow rate and tangential velocity. For pump flow output the second highest parameter is no. of pumps installed in the well, and for tangential velocity the second highest parameter is pump inlet bell velocity. In the present study of pump sump optimization pump, c/c distance

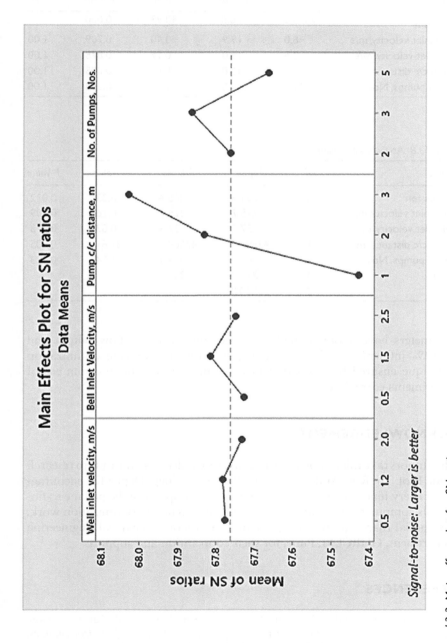

Figure 13.3 Main effect plot for SN ratios.

Table 13.7 Coefficients

Term	Coef	SE Coef	T-Value	P-Value	VIF
Constant	2323.2	56.8	40.93	0.000	
Well inlet velocity, m/s	−8.0	19.9	−0.40	0.709	1.00
Bell inlet velocity, m/s	2.5	15.0	0.17	0.875	1.00
Pump c/c distance, m	84.2	15.0	5.62	0.005	1.00
No. of pumps, Nos.	−12.14	9.80	−1.24	0.283	1.00

Table 13.8 Analysis of variance

Source	DF	Adj SS	Adj MS	F-Value	P-Value
Regression	4	44821.6	11205.4	8.33	0.032
Well inlet velocity, m/s	1	215.7	215.7	0.16	0.709
Bell inlet velocity, m/s	1	37.5	37.5	0.03	0.875
Pump c/c distance, m	1	42504.2	42504.2	31.61	0.005
No. of pumps, Nos.	1	2064.3	2064.3	1.54	0.283
Error	4	5378.4	1344.6		
Total	8	50200.0			

parameters have about 84.66% influence on the pump flow output and 71.72% influence on the tangential pump velocity. Taguchi optimization technique ensured the reduction in experimentation time and number of experiments carried out.

ACKNOWLEDGEMENTS

The authors take this opportunity to convey their deepest gratitude to research guide Prof. Dr. R.R. Arakerimath (HOD, Mech. Engg. Dept.) for encouraging at every juncture of need. His support and cooperation helped me to finish this computational analysis work and Taguchi design optimization work. My special thanks to all post graduate faculties of Mechanical Engineering Department, GHRCEM, Pune, for their cooperation and support.

REFERENCES

1. C. Hong Xun and G. Jia-Hong, "Numerical Simulation of 3D Turbulent Flow in the Multi Intakes Sump of the Pump Station", *Journal of Hydrodynamics*, 19–1, 42–47, 2007. DOI: 10.1016/S1001-6058(07)60026-2

2. A. Bayeul-Lainé, G. Bois and A. Isa, "Numerical Simulation of Flow Field in Water-Pump Sump and Inlet Suction Pipe", *25th IAHR Symposium on Hydraulic Machinery and Systems, IOP Conf. Series: Earth and Environmental Science*, 12, 2010. DOI: 10.1088/1755-1315/12/1/012083

3. T. Constantinescu, "Validation of a Large-Eddy Simulation Model to Simulate Flow in Pump Intakes of Realistic Geometry", *Journal of Hydraulic Engineering*, 132, 1303–1315, 2006. DOI: 10.1061/(ASCE)0733-9429(2006)132:12(1303)

4. T. Kueh, S. Beh, D. Rilling and Y. Ooi, "Numerical Analysis of Water Vortex Formation for the Water Vortex Power Plant", *International Journal of Innovation Management and Technology*, 5(2), 111–115, 2014. DOI: 10.24084/repqj18.259

5. A. Amin, B. Ha Kim and C. Gu Kim, "Numerical Analysis of Vortices Behavior in a Pump Sump", *IOP Conference Series Earth and Environmental Science*, 240(3), 2019. DOI: 10.1088/1755-1315/240/3/032020

6. ANSI/HI 9.8.2012, American National Standard for Rotodynamic Pumps for Pumps Intake Design in the *Hydraulic Institute*, New Jersey.

7. George Constantinescu and Virendra Patel, "Numerical Model for Simulation of Pump-Intake Flow and Vortices", *Journal of Hydraulic Engineering*, 124, 123–134, 1998. DOI: 10.1061/(ASCE)0733-9429(1998)124:2(123).

8. G. Taguchi, S. Chowdhury and Y. Wu, (2005). *Taguchi's Quality Engineering Handbook*. John Wiley & Sons, New Jersey.

9. Y. Kazançoğlu and M. Bayramoglu, "Multi-Objective Optimization of the Cutting Forces in Turning Operations Using the Grey-Based Taguchi Method", *Materials and Technologies*, 45(2), 105–110, ISSN 1580-2949, 2011.

10. J. Arnott and Bryan, "Optimising Pump Selection in Sweden", *World Pumps* (477), 40–43, 2006. DOI: 10.1016/S0262-1762(06)70985-3

11. Y. Zhao, B. Liang and Yang Wang, "Optimising Pump Scheduling for Water Distribution Networks", *Advances in Artificial Intelligence*, 2019. DOI: 10.1007/978-3-030-35288-2_35

12. M. Şeker, I. Mutlu and F. Emre Aysal, "The ANN Analysis and Taguchi Method optimisation of the Brake Pad Composition", *Emerging Materials Research*, 10(3), 314–320, 2021. DOI: 10.1680/jemmr.21.00036

Chapter 14

Prediction of fire-resistance rate of CFST columns using gene expression programming

Aishwarya Narang
Center of Excellence in Disaster Mitigation & Management,
Indian Institute of Technology Roorkee, Roorkee, India

Ravi Kumar and Amit Kumar Dhiman
Indian Institute of Technology Roorkee, Roorkee, India

CONTENTS

INTRODUCTION

As more high-rise contemporary structures are built worldwide, structural fire safety is becoming an increasingly significant part of their design. Concrete-filled steel tubes (CFST) have become more prevalent in recent decades as a result of several advantages [1–3]. Ease of construction with increased stiffness and ductility, comparatively smaller cross-sectional area than conventional RCC columns with a higher load-bearing capacity, provides better resistance to local buckling in CFST columns [4, 5]. Furthermore, without the need for external protection, this structural component can attain a high level of fire resistance. Steel degrades at temperatures beyond 600°C during a fire, and steel columns without a concrete core are vulnerable [6, 7]. The concrete core's combined action with the steel tube results in outstanding fire resistance. Because of the benefits of CFST columns and their growing popularity, a decisive approach is required to estimate the fire-resistance performance of the CFST columns during the design phase and select the most influential parameters to increase the fire resistance [8]. Furthermore, understanding the degree of load-bearing capacity loss in the CFST column during fires over a specific time period will assist engineers in providing

DOI: 10.1201/9781003257691-14

187

more exact and informative ways for reinforcing damaged CFST columns. Furthermore, the limitations of empirical approaches and the design code linkages highlight the need for more research in this area.

A machine learning methodology is used to analyze the extensive existing CFST columns database to produce simple analysis and design tools. In this study, a simple yet reliable model based on GEP is developed to estimate the fire resistance of CFST columns. According to this database, an extensive trial and error procedure was used to select the best GEP model with the lowest mean square error (MSE) and the highest correlation coefficient (R). Gene expression programming (GEP) is a genetic algorithm (GA) in the same way as GAs and genetic programming (GP) are. It takes populations of people, chooses them based on fitness, and adds genetic diversity using one or more genetic operators [9].

GEP is a prominent soft computing technology used by a variety of researchers in various engineering fields. The replication of DNA molecules at the gene level is the source of actual GEP. GEP, a version of GP that uses fixed-length linear chromosomes and encodes a brief program, was recently developed [9]. GEP offers the benefit of being able to depict the outcome using a simple mathematical expression that is acceptable for practical use and improved forecast accuracy.

DATA COLLECTION

For the prediction of an expression programming to compute the FRR of CFST columns, a detailed literature survey was performed. Non-protected CFST columns were considered for the study. There is no consideration for double skin CSFT columns in the database. Considering the literature review, seven parameters that influence the fire performance of these columns were selected [8]. Concrete compressive strength (f_c), Yield Strength of steel (f_y), Cross-sectional area (A_c), Slenderness ratio (λ), Thickness of steel (t_s), Load ratio (n), and Eccentricity (e) were used as the input parameters for the CFST columns. 182 data points were collected from the literature [1–7, 10–20]. Out of 182, 122 datasets were taken for the testing purpose and the rest to validate the achieved model. Table 14.1 has the list of input parameters and their statistical properties. Also, Figure 14.1 (a–h) shows the data distribution curve for all the collected data (182 columns). From the database, it can be said that the data has a wide range of values.

RESEARCH METHODOLOGY

GEP is a prominent soft computing technology used by a range of researchers in various engineering fields. At the gene level, GEP is created by the replication of DNA molecules. Leads to activation of non-linear parse tree-like

Table 14.1 Statistical properties of input parameters

Input parameters	Minimum	Maximum	Average	Median
d0: f_c	23.8	101.6	51.98	41.935
d1: A_c	15614.04	179359.94	47212.70	37675.57
d2: λ	6.58	94	46.170	46.16
d3: t_s	2.95	12.7	6.612	6
d4: f_y	292.9	569	370.67	350
d5: n	0.09	0.75	0.337	0.33
d6: e	0	0.75	0.0053	0

architectures, GEP is an adaptive programming paradigm. Depending on the data, it estimates appropriate starting non-linearity. The GEP's execution time is determined by the chromosomal level, which determines the population size [9]. The chromosome that produces the greatest outcomes is passed down to reproductive success, and the cycle is continued until an acceptable level of fitness is achieved.

The essential data for constructing an empirical relationship is encoded mostly in chromosomes, and a MATLAB program is also created to infer this data. Multiplication, addition, division, and subtraction functions are used to develop the expression tree and the required equations. Multiplication is used to interlink the functions in GEP model. Figure 14.2 shows the flow chart for the GEP modeling for a better understanding.

RESULTS AND DISCUSSION

GEP delivers the solution for predicting a parameter in the form of expression trees. From these expression trees, the equation for the FRR can be achieved. GeneXprotools 5.0 is used to generate the GEP model. The expression tree diagram for the observed data is shown in Figure 14.3, while Figure 14.4 shows the variation between the predicted values and experimental values of FRR.

A correlation coefficient is a measure of the performance of any model. The observed values of the correlation coefficient are 0.802 and 0.779 for the testing and validation database, respectively. These values show that the GEP can predict the FRR for low to high strength concrete with different eccentric load ratios.

The existing literature data are in a wide range of variety, and the correlation results of data frequency are not well suited. The complexity of testing the CFST columns with asymmetric loading can be assumed the reason behind the poor correlation. Despite that, GEP gives a fair agreement between the predicted (model data) and experimental (target data) values of FRR.

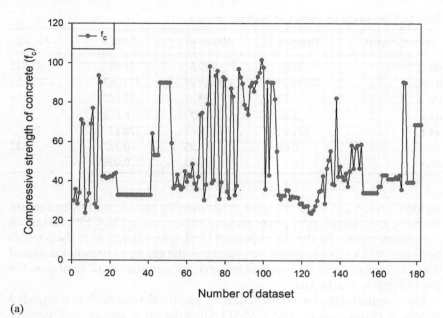

(a)

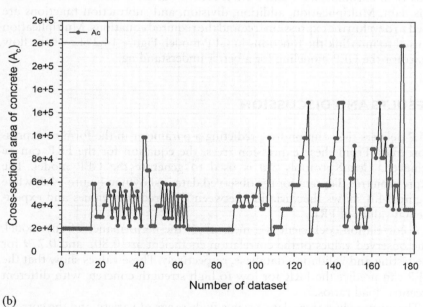

(b)

Figure 14.1 Frequency distribution curve for the selected parameters: (a) Compressive strength of concrete (MPa); (b) Cross-sectional area of concrete (mm²);

(*Continued*)

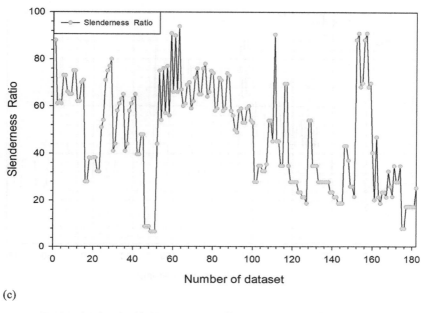

(c)

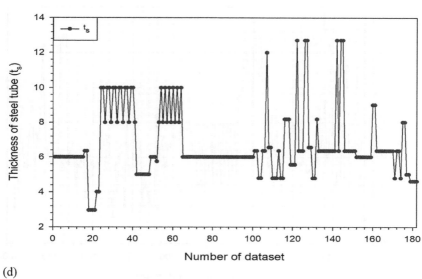

(d)

Figure 14.1 (Continued) (c) Slenderness ratio; (d) Thickness of steel tube (mm);

(Continued)

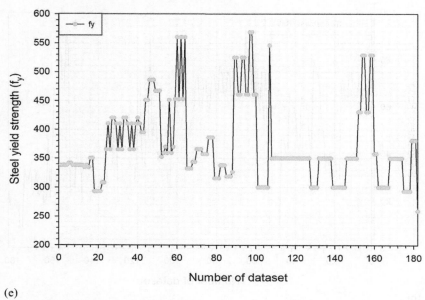

(e)

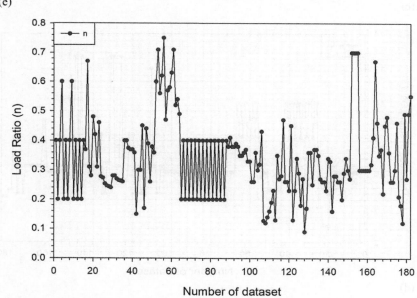

(f)

Figure 14.1 (Continued) (e) Steel yield strength (MPa); (f) Load ratio.

(*Continued*)

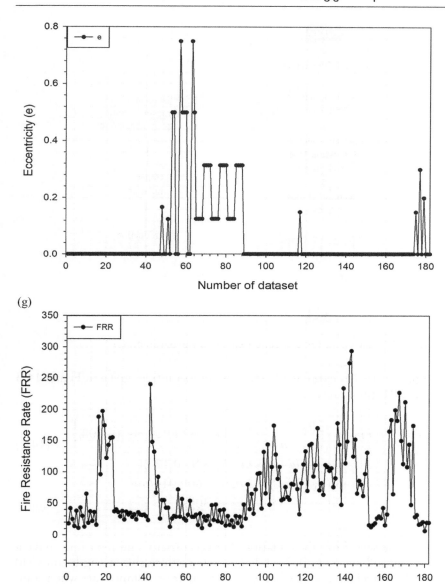

(g)

(h)

Figure 14.1 (Continued) (g); Eccentricity; (h) Fire- resistance Rate (in minutes).

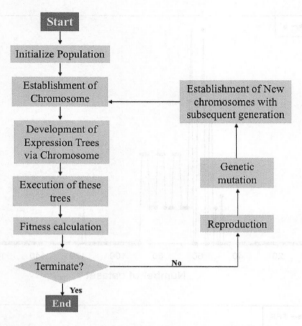

Figure 14.2 Flow chart for the gene expression programming.

Equation 14.1 is extracted from the expression trees given in Figure 14.3 for the FRR (in minutes).

$$FRR(in\ mins) = \left\{ \left(\frac{(ts - 7.965) \times ts}{n \times \lambda} \right) + ((-32.692 \times e) + 16.708) \right\}$$

$$\{(e - (((e - ts) \times n) + 5.752) \times 0.120)\}$$
$$\{(((((Acl - 0.363)/ts) + (-609.010/n))/$$
$$((fy \times 6.785) + (fc \times -9.308))\}$$

(14.1)

It can be seen from the equation that eccentricity and load ratio have a significant influence on FRR while concrete compressive strength and yield strength of steel have somewhat less effect on fire-resistance rate when compared. The sensitivity of the target and model can also be seen in Figure 14.4, where there is a significant agreement between the two values.

CONCLUSIONS

The gene expression programming methodology (GEP) is used in this study to create an expression for measuring the FRR of CFST columns. The suggested GEP model is based on a broadly distributed database of various

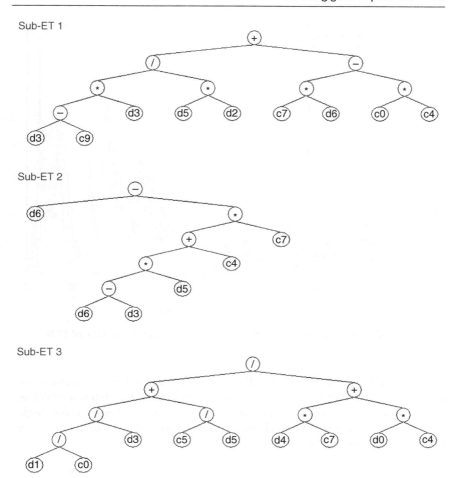

Figure 14.3 Expression tree diagram for GEP model.

parameters obtained from existing published literature. After examining the results of testing data and graphs, it can be concluded that due to significant variation in the dataset, the regression is observed as 0.802. This is because all the data were taken from the existing literature and comprised different concrete strength and testing conditions.

Using GEP programming generates equations that can predict fire performance of CFST columns when there is not a specific standard to select eccentric loading method on low- to high-performance concrete. According to the parametric analysis, the projected model correctly incorporates the effects of the input parameters to anticipate the precise structure of FRR. The correctness of the projected models is checked using the correlation coefficient and RMSE for training samples. Furthermore, the model satisfies all of the necessary criteria for validation data as well.

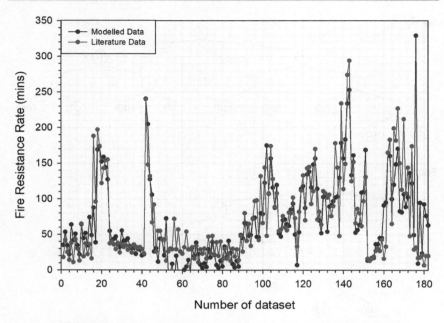

Figure 14.4 Comparison graph for the modeled and literature data of FRR.

The projected GEP model can offer a thorough as well as pragmatic foundation for using CFST columns for construction practices while protecting them from fire hazards. The research gives an exact idea of using high-performance concrete and steel encasing in CFST columns to increase the FRR to cope with any disaster.

NOMENCLATURE

A_c	Cross-sectional area	[mm²]
f_c	Compressive strength of concrete	[MPa]
f_y	Yield strength of steel	[MPa]
n	Load ratio	–
e	Eccentricity	–
λ	Slenderness ratio	–
t_s	Thickness of steel	[mm]

REFERENCES

1. M.L. Romero, V. Moliner, A. Espinos, C. Iba˜nez, A. Hospitaler, "Fire behavior of axially loaded slender high strength concrete-filled tubular columns," *J. Constr. Steel Res.* 67 (12) (2011) 1953–1965. https://doi.org/10.1016/j.jcsr.2013.01.011

2. Z. Tao, M. Ghannam, T.Y. Song, L.H. Han, "Experimental and numerical investigation of concrete-filled stainless steel columns exposed to fire," *J. Constr. Steel Res.* 118 (2016) 120–134. https://doi.org/10.1016/j.jcsr.2015.11.003

3. V. Moliner, A. Espinos, M.L. Romero, A. Hospitaler, "Fire behavior of eccentrically loaded slender high strength concrete-filled tubular columns," *J. Constr. Steel Res.* 83 (2013) 137–146. https://doi.org/10.1016/j.jcsr.2013.01.011

4. J. Rodrigues, L. Laim, "Fire response of restrained composite columns made with concrete filled hollow sections under different end-support conditions," *Eng. Struct.* 141 (2017) 83–96. https://doi.org/10.1016/j.engstruct.2017.02.073

5. L.H. Han, F. Chen, F.Y. Liao, Z. Tao, B. Uy, "Fire performance of concrete filled stainless steel tubular columns," *Eng. Struct.* 56 (2013) 165–181. https://doi.org/10.1016/j.engstruct.2013.05.005

6. H. Yang, F. Liu, L. Gardner, "Performance of concrete-filled RHS columns exposed to fire on 3 sides," *Eng. Struct.* 56 (2013) 1986–2004. https://doi.org/10.1016/j.engstruct.2013.08.019

7. L.H. Han, Y.F. Yang, L. Xu, "An experimental study and calculation on the fire resistance of concrete- filled SHS and RHS columns," *J. Constr. Steel Res.* 59 (4) (2003) 427–452. https://doi.org/10.1016/S0143-974X(02)00041-X

8. M.J. Moradi, K. Daneshvar, D. Ghazi-Nader, H. Hajiloo, "The prediction of fire performance of concrete-filled steel tubes (CFST) using artificial neural network," *Thin-Walled Structures*, (2021). https://doi.org/10.1016/j.tws.2021.107499

9. M.A. Khan, A. Zafar, A. Akbar, M. Javed, A. Mosavi. "Application of Gene Expression Programming (GEP) for the prediction of compressive strength of geopolymer concrete," *Materials*, (2021). https://doi.org/10.3390/ma14051106

10. K.U. Ukanwa, J.B. Lim, U.K. Sharma, S.J. Hicks, A. Abu, G.C. Clifton, "Behaviour of continuous concrete filled steel tubular columns loaded eccentrically in fire," *J. Constr. Steel Res.* 139 (2017) 280–287. https://doi.org/10.1016/j.jcsr.2017.09.030

11. N. Tondini, V.L. Hoang, J. Demonceau, J. Franssen, "Experimental and numerical investigation of high- strength steel circular columns subjected to fire," *J. Constr. Steel Res.* 80 (2013) 57–81. https://doi.org/10.1016/j.jcsr.2012.09.001

12. T. Lie, "Fire resistance of circular steel columns filled with bar-reinforced concrete," *J. Struct. Eng.* 120 (5) (1994) 1489–1509. https://doi.org/10.1061/(ASCE)0733-9445(1994)120:5(1489)

13. V. Kodur, "Solutions for enhancing the fire endurance of HSS columns filled," *Eng. J.* 43 (1) (2006) 1–7.

14. A. Espinos, M.L. Romero, E. Serra, A. Hospitaler, "Circular and square slender concrete-filled tubular columns under large eccentricities and fire," *J. Constr. Steel Res.* 110 (2015) 90–100. https://doi.org/10.1016/j.jcsr.2015.03.011

15. H. Yang, F. Liu, L. Gardner, "Performance of concrete-filled RHS columns exposed to fire on 3 sides," *Eng. Struct.* 56 (2013) 1986–2004. https://doi.org/10.1016/j.engstruct.2013.08.019

16. L.H. Han, Y.F. Yang, L. Xu, "An experimental study and calculation on the fire resistance of concrete- filled SHS and RHS columns," *J. Constr. Steel Res.* 59 (4) (2003) 427–452. https://doi.org/10.1016/S0143-974X(02)00041-X

17. H. Lu, X.L. Zhao, L.H. Han, "Fire behaviour of high strength self-consolidating concrete filled steel tubular stub columns," *J. Constr. Steel Res.* 65 (10–11) (2009) 1995–2010. https://doi.org/10.1016/j.jcsr.2009.06.013

18. T.A. Pires, J.P.C. Rodrigues, J.J.R. Silva, "Fire resistance of concrete filled circular hollow columns with restrained thermal elongation," *J. Constr. Steel Res.* 77 (2012) 82–94. https://doi.org/10.1016/j.jcsr.2012.03.028

19. V. Kodur, T.T. Lie, "Experimental studies on the fire resistance of circular hollow steel columns filled with steel fibre reinforced concrete," National Research Council Canada, Institute for Research in Construction, 1995. https://doi.org/10.4224/20375221

20. V.K.R. Kodur, "Performance of high strength concrete-filled steel columns exposed to fire." *Can. J. Civ. Eng.* 25 (6) (1998) 975–981.

Chapter 15

Computational comparative investigation of baseline and perforated pin-fins

Mohak Gaur and Amit Arora

Malaviya National Institute of Technology Jaipur, Rajasthan, India

CONTENTS

15.1 INTRODUCTION

The development of integrated circuits and CPUs has increased their computational power, but it has also resulted in an increase of the heat generated by them and with the current demand of miniaturizing electronic devices, the space available to dissipate the heat generated is limited. Therefore, the need is to design heat sinks that can dissipate more heat in the limited space available. For the effective working of processors, the operating temperature should be maintained in the 60°C–80°C range. A good design of heat sink is the one which has high heat transfer rate and low-pressure loss/drop across the system. There are several key factors, which should be considered while designing the heat sink like operating temperature range, mass of heat sink, heat sink conductivity, flow condition of working fluid, cost of fabrication volume, and dimensions of the heat sink. Until recently, research was focused on using simplified models for numerical analysis of heat sinks, but with the development in technology and the rising capabilities of computational fluid dynamics (CFD) simulations, it is possible to conduct numerical analysis for thermal and pressure loss characteristics of complex heat sink geometries. A number of techniques have been employed to increase the heat transfer rate and reduce the pressure loss/drop across heat sink. The work methodologies can be of three types, using nanofluids, geometric modification, or adopting non-conventional cooling techniques.

Sohel et al. [1] conducted experiments that have shown that bulk increase in Al_2O_3/H_2O nanofluid concentration in the range of 0.1%–0.25% resulted in increase of thermal efficiency, but thermal efficiency did not necessarily increase with the increase in flow rate. It was found that the convective heat transfer coefficient increases by 18% using 0.25% concentrated Al_2O_3/H_2O nanofluids compared to when distilled water was used. Singh et al. [2] demonstrated the effect of using nanoparticles as working fluid on thermal performance of heat sink. Three volume concentrations of 0.25%, 0.5%, and 1.0% Al_2O_3 nanoparticles in a base fluid of water and ethylene glycol were used and two different nanofluids with particle size of 45 nm and 150 nm were used. The working fluid was passed through microchannels having a hydraulic diameter D_h of 130 μm, 211 μm, and 300 μm. It was observed that an early transition to turbulent flow was taking place for channels having higher hydraulic diameter. It was attributed to the high surface roughness of these microchannels. Anoop et al. [3] used three weight concentrations, 0.2%, 0.5%, 1% of SiO_2/H_2O nanofluids and analyzed their effect on heat transfer enhancement. A poly di-methyl siloxane microchannel was used by them as the heat sink. The Reynolds number was varied from 4000 to 22,000. They found that nanofluids had higher heat dissipation rate at lower flow velocity.

An experimental study was conducted to estimate the effectiveness of inline pin fin array and staggered pin fin array by Sparrow et al. [4]. Two main factors considered by them to measure effectiveness were heat dissipation and the pressure drop across the heat sink. They found out that the staggered configuration of pin fin array was better than inline configuration of pin fin array for the purpose of heat dissipation. In addition, lower pressure drop across the heat sink was observed for the staggered configuration. It was concluded from this study that the inline pin fins configuration has better thermal dissipation capability as compared to staggered pin fins configuration if equal pumping power and equal surface area were taken for both cases, while the staggered pin fin configuration requires less heat transfer area than inline pin fin configuration for a fixed heat flux supplied and a fixed mass flow rate of working fluid. Bilen et al. [5–7] studied experimentally the friction loss as well as the heat transfer characteristics of pin fin heat sink of both staggered and inline configuration of pin fins in a rectangular channel. It was reported that both staggered pin fin and inline configuration performed better than plane heat sink (no projected surfaces/pin fins attached). Correlations for friction factor and Nusselt number was developed through these experiments. It was concluded that at a constant pumping power and equivalent Reynolds number, the staggered configuration of pin fins had heat transfer enhancement of 33% compared to that of plane heat sink. Ismail et al. [8] conducted a numerical study on the thermal performance of perforated pin fin heat sink having different cross-section of perforations. The shapes of perforated channel cross-section analyzed were square, circular, triangular, and hexagonal. It was found that for the same surface area circular perforated fins had greater fin effectiveness value (PFE) as well as the lower drag force compared to the other shape of perforations. Saadat et al. [9] conducted numerical analysis of turbulent convection heat transfer from pin fins having cross perforations. It was found that the pin fins having one longitudinal and three vertical perforations had the highest heat transfer performance. Ismail et al. [10] numerically investigated the convective heat transfer from turbulent flow of air over solid rectangular fins having longitudinal perforations. It was found that the fins having circular perforation had substantial heat transfer enhancement as well as reduced pressure drop across heat sink compared to other geometries.

Naphon et al. [11] analyzed the effect of using thermoelectric on the thermal performance of liquid cooled CPU. It was found that using water as working fluid along with thermoelectric to cool the CPU resulted in reduction of operating temperature but an increase in energy consumption was also observed. The increase in energy consumption was attributed to the additional energy required to operate the thermoelectric. Hu et al. [12] conducted an experimental study on the effect of using water-cooled thermoelectric on the thermal performance of CPU operated under severe environmental condition. It was found that the largest fluctuation in operating temperature T_{cpu} with temperature control under variable heat flux was less than 1.4°C. They concluded that the passive cooling techniques are not feasible in severe environmental conditions. It was found that using thermoelectric with water-cooled heat sink along with a temperature controller can prevent dew and overheating under variable operating conditions as well as reduce the energy consumption and thus was found out to be more suitable for processors operating under harsh environmental conditions.

15.2 COMPUTATIONAL MODELING

15.2.1 Computational domain and boundary conditions

The staggered perforated pin fin heat sink has been simulated using Ansys 15.0-Fluent software. The flow was considered to be three-dimensional incompressible and had achieved steady state. Dimensions of the fluid region of computational domain as well as the boundary conditions were based on the experimental setup of the experiment performed by Chin et al. [13]. SIMPLEC algorithm was employed to couple velocity and pressure. Finite volume method was employed. The equations that were solved to get solution are namely the Energy Equation, Navier–Stokes equations, and continuity equation. k-Epsilon Realizable model with the standard wall function was used to simulate the turbulent flow in channel. The equations can be referred to from Ref. [14]. The boundary conditions at inlet were taken to be velocity inlet with uniform velocity, temperature of 300K, and atmospheric pressure (Gauge pressure = 0). Turbulence intensity was taken to be 10% as it was in the experiment and the hydraulic diameter was calculated to be $D_h = 0.067$ m. The fluid domain walls and pin fin walls had no-slip conditions. From downstream of the heat sink to the outlet, the flow was fully developed. The fluid domain walls were adiabatic, excluding the base plate on which a constant heat flux of 5903 W/m² was given.

The fluid domain was constructed as shown in Figure 15.1. The test section length was 100 mm while the entrance length was 800 mm and downstream to exit length was 200 mm. The lengths of sections were taken as such so that the flow was fully developed upstream and downstream the heat sink. The pressure drop was calculated by taking the difference of the average surface value of pressure on a plane 10 mm upstream and 10 mm downstream as was the position of pressure gauge in the experiment. All the walls of fluid domain were taken as adiabatic and a constant uniform heat flux was given to the base plate of heat sink. Velocity inlet boundary condition was taken at inlet while outflow boundary condition was given to the outlet. The readings were taken at four different Reynolds number 8711.91, 13199.4, 17576.2, 22063.7, and the corresponding inlet velocities were 1.89 m/s, 2.88 m/s, 3.83 m/s, and 4.81 m/s, respectively.

The computational domain was discretized into grids using cutcell method. The number of control volumes for the solid pin fin array was approximately 1×10^6. Finer mesh structures were located in the vicinity of the fins and the perforations as can be seen from Figure 15.2. Grid

Boundary No.	Name	Type
1	Wall	Adiabatic
2	Outlet	Outflow
3	Inlet	Velocity
4	Base plate	Heat flux

Note: All dimensions are in mm

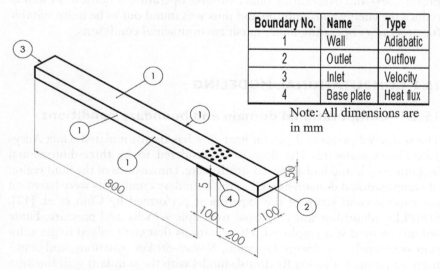

Figure 15.1 Fluid region of computational domain.

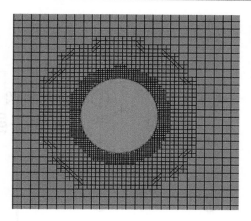

Figure 15.2 Meshing of fluid region around the pin fin.

independence test was conducted for solid pin fin array, a 30% increase in number of elements changed average Nusselt number by 1.4% and the average pressure drop across the heat sinks by about 0.6%.

The formula used to calculate various parameters are given below:

$$Nusselt\ number\,(Nu) = \frac{q_{00}D_h}{k_{air}\left(T_w - \dfrac{T_{out} + T_{in}}{2}\right)} \tag{15.1}$$

$$Pressure\ drop\,(\Delta P) = P_{in} - P_{out} \tag{15.2}$$

$$System\ performance\,(\eta) = \frac{Nu}{\dfrac{\Delta P}{0.5\rho u_0{}^2}} \tag{15.3}$$

$$Reynolds\ number\,(Re_h) = \frac{\rho u_0 D_h}{\mu} \tag{15.4}$$

$$Hydraulic\ diameter\ (D_h) = \frac{4A}{P_c} \tag{15.5}$$

15.2.2 Heat sink dimensions

The staggered pin fin arrangement was taken on a rectangular 100 mm × 100 mm base plate containing 14 circular pin fins. The diameter of each pin fin was 8 mm. The arrangement of pin fin is as shown in Figure 15.3. Four different configurations of pin fins were analyzed, including one solid pin fin geometry and three perforated geometries.

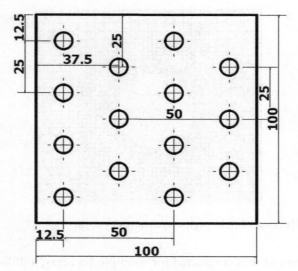

Note: All dimensions are in mm

Figure 15.3 Heat sink configuration.

Note: All dimensions are in mm.

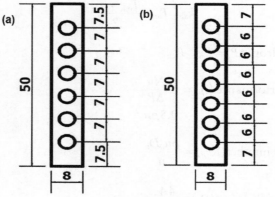

Note: All dimensions are in mm

Figure 15.4 Perforated pins (a) P_p = 7 mm (b) P_p = 6 mm.

Note: All dimensions are in mm.

The configuration of perforations on the pin fins is as shown in Figure 15.4, for each configuration the perforation diameter was varied as 2 mm, 2.5 mm, 3 mm, 3.5 mm, and 4 mm. To verify the consistency of results, the results were calculated for perforation pitch of 7 mm and 6 mm.

15.3 MODEL VALIDATION

The experimental data were taken from the experiment conducted by Chin et al. [13]. The computational model dimensions as well as upstream and downstream section lengths were taken similar to that of the experimental model. The heat sink material was aluminum, while the working fluid was air. Figure 15.5 depicts the contrast between the experimentally measured and the simulated Nusselt number and pressure loss across the test section against Re_h for heat sink with solid pin fins. As can be seen from the graph, both Nu and ΔP increase with the increase in Re_h. The computationally calculated pressure loss across heat sink is slightly over estimated as

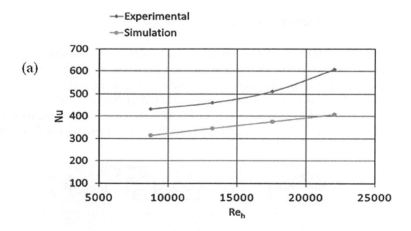

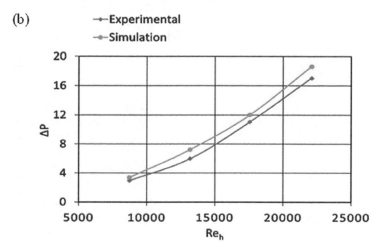

Figure 15.5 Model validation (a) Nusselt number (b) Pressure drop.

compared to the experimentally determined pressure drop; at low value of Reynolds number, pressure drop was 14.64% more than the experimental value, while Nusselt number was 27.34% lower than the experimental value. This could be because no clearance is given between fin top and the top wall of computational domain, while such perfect contact is possible in simulation, there might be imperfect contact between the top of pin fins and the top wall of fluid domain, causing gaps between them in the experimental setup which may allow bypass flow leading to reduced pressure drop. The limitations of k-epsilon realizable turbulence model are also a contributory factor in the divergence of Nusselt number from experimental data at higher Reynolds number (Re_h).

15.4 RESULTS AND DISCUSSION

In order to appreciate the changes in flow characteristics, streamlines need to be analyzed. For that purpose, fins with least perforation pitch (i.e., 6 mm) and highest perforation diameter (i.e., 4 mm) are considered, as shown in Figure 15.6. The comparative study is conducted at $Re_h = 22063.7$. As evident, the perforated fins show better flow characteristics compared to solid fins because perforated pin fins give clear path to the approaching flow and

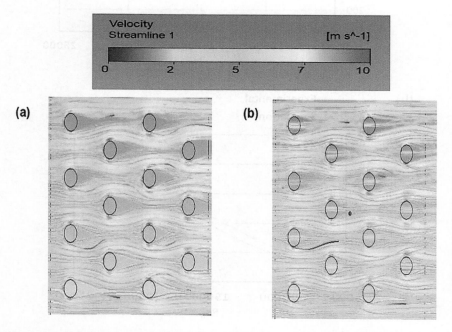

Figure 15.6 Streamlines (a) Solid Pin Fin (b) Perforated Pin fins with P_p = 6 mm and D_p = 4 mm.

thus the approaching fluid experiences lesser obstruction and also the contact area between fluid and the pin fins increases, which is the reason for the enhanced thermal performance.

Due to such geometric modifications, streamlines are changed around the pin fin array as depicted in Figure 15.6. Furthermore, it can be seen that the streamlines bend around the solid fins while they pass through the perforations provided in the modified design, and thus lesser obstruction is experienced by the air, which helps in lowering the pressure drop. In addition, narrower and smaller recirculation region behind the perforated fins facilitates greater heat transfer augmentation. Various local flow modifications are going to manifest as there is an increase in average Nusselt number as well as pressure loss. The discussion on change in various average performance parameters is discussed in the subsequent sub-sections.

15.4.1 Change in hydraulic performance

A graph between the pressure loss/drop across the heat sink against the Reynolds number for different diameter of perforation D_p for a defined perforation pitch P_p has been plotted (Figure 15.7). It can be clearly seen that the pressure drop/loss across the heat sink increases with the increase in Reynolds number. It can also be seen that the pressure drop reduces with the increase in diameter of perforation for a specific pitch of perforations.

For a perforation pitch of P_p = 7 mm, the pressure loss for perforation diameter of 4 mm (D_p = 4 mm) was 11.97% lower at low Reynolds number and 30.16% lower at high Reynolds number when compared with the pressure drop of configuration having perforation diameter of 2 mm (D_p = 2 mm).

A similar trend was noticed when the perforation pitch was set at a value of 6 mm (P_p = 6 mm). The pressure loss/drop reduces with the increase in perforation diameter. The pressure loss for perforation diameter of 4 mm (D_p = 4 mm) was 13.93% lower at low Reynolds number and 34.65% lower at high Reynolds number when compared with the pressure drop configuration having a perforation diameter of 2 mm (D_p = 2 mm).

15.4.2 Change in thermal performance

Graphs have been plotted for the Nusselt number against the Reynolds number for different perforation diameter D_p for a preset value of perforation pitch P_p. It has been noticed that the Nusselt number increases initially with increase in perforation diameter D_p but on further increasing the perforation diameter from 3 mm results in either a marginal increase or a marginal reduction as depicted in Figure 15.8. For the perforation pitch of P_p = 7 mm at high value of Reynolds number on increasing D_p from 3 mm to 3.5 mm resulted in an increase in Nusselt number by 0.066% while on further increasing D_p from 3.5 mm to 4 mm, it resulted in an increase

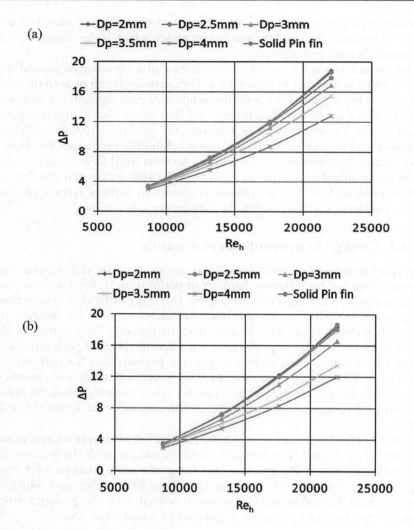

Figure 15.7 Effect of perforation diameter on pressure drop (a) P_p = 7 mm
(b) P_p = 6 mm.

of Nusselt number by 0.28%. Similarly for the perforation pitch of 6 mm, dominance of configuration with D_p = 3.5 mm at high Reynolds number was observed as it resulted in improvement of Nusselt number by 45.41% when compared to that of solid pin fins under similar operating conditions. On further increasing perforation diameter from 3.5 mm to 4 mm resulted in reduction in Nusselt number by 2.08%. This might be the result of the counteracting effects of increase in convectional heat transfer but a reduction in conductive heat transfer along the pin fins axis.

(a)

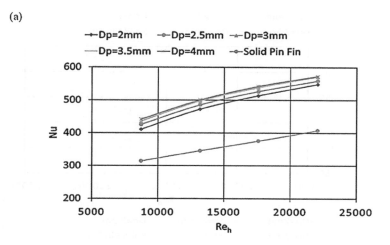

(b)

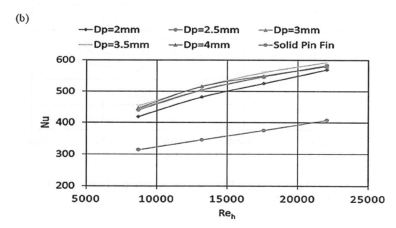

Figure 15.8 Effect of perforation diameter on Nusselt number (a) P_p = 7 mm (b) P_p = 6 mm.

15.4.3 Change in system performance 'η'

The system performance η of a heat sink is a dimensionless number defined as the ratio of Nusselt number to the pressure coefficient. This parameter can be used in order to determine the optimum design based on the compromise between most heat dissipation rate and the least pressure loss/drop across heat sink as it determines the relative cost, i.e., pumping loss/pressure drop to attain a definite amount of heat transfer.

As can be inferred from Figure 15.9, the system performance increased with the increase in perforation diameter for the perforation pitch of 7 mm as well as for perforation pitch of 6 mm. For the case of P_p = 7 mm, the perforation diameter of 4 mm gave the best system performance for all flow

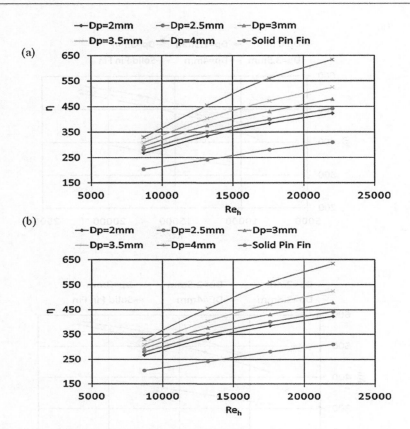

Figure 15.9 Effect of perforation diameter on system performance (a) P_p = 7 mm (b) P_p = 6 mm.

velocities. The η for P_p = 7 mm and Dp = 4 mm was 2.04 times of what it was for solid pin fins at an inlet velocity of 4.81 m/s. A similar trend was followed by configuration with P_p = 6 mm. The perforation diameter of 4 mm gave better system performance for all flow velocities. The η for configuration with P_p = 6 mm and D_p = 4 mm was 2.21 times of what it was for solid pin fins at an inlet velocity of 4.81 m/s.

15.5 CONCLUSION

Steady and incompressible flow of air over staggered perforated pin fin array in a rectangular channel has been analyzed computationally to calculate their convective heat transfer coefficient and pressure drop. Hydraulic and thermal performance of various perforated pin fin geometries have been compared and studied, specifically the effect of perforation diameter on system performance has been studied. In order to examine the effect of perforation size, five different perforation diameters are considered, and their performance is compared with the solid pin fins over a wide range of Reynolds number ranging from 8000 to 22,000. The consistency of trends of this study is analyzed by considering two different perforation pitches. The conclusions of this study are:

- The hydraulic performance of the heat sink improved with the increase in perforation diameter. Among all the configuration of perforated pin fins analyzed, all of them had better pressure drop characteristics as compared to their solid counterpart.
- Nusselt number initially increased with the increase in perforation diameter but increasing the perforation diameter beyond 3 mm results in either a marginal increase or a marginal reduction in the value of Nusselt number. This might be the result of counteracting effects of increase in heat transfer by convective mode but a reduction in conductive heat transfer along the pin fins axis.
- The system performance of both the selected perforation pitch increased with the increase in perforation diameter. All the perforated configurations had a better system performance when compared to that of solid pin fins. The system performance for the configuration $P_p = 6$ mm and $D_p = 4$ mm was 2.21 times of what it was for solid pin fins at high Reynolds number.

NOMENCLATURE

D	Pin diameter	[mm]
D_h	Hydraulic diameter of rectangular channel	[mm]
D_p	Diameter of perforation	[mm]
k_{air}	Thermal conductivity of air	[W m^{-1} K^{-1}]
ρ	Density of air	[kg/m^3]
μ	Viscosity of air	[Kg m^{-1} s^{-1}]
Nu	Nusselt number	–
ΔP	Pressure loss across heat sink	[Pa]
η	System performance	–
η_{ratio}	Performance ratio	–
Re_h	Reynolds number	–
u_o	Inlet velocity	[ms^{-1}]
T_{in}	Inlet temperature	[°C]
T_{out}	Outlet temperature	[°C]
T_w	Average base plate temperature	[°C]

D	Pin diameter	[mm]
q_{00}	Heat flux	[Wm^{-2}]
P_p	Perforation pitch	[mm]

REFERENCES

1. Z. Khattak, H.M. Ali, "Air cooled heat sink geometries subjected to forced flow: A critical review", *International Journal of Heat and Mass Transfer*, vol. 130, pp. 141–161, 2019. https://doi.org/10.1016/j.ijheatmasstransfer.2018.08.048
2. M. Nazari, M. Karami, M. Ashouri, "Comparing the thermal performance of water, Ethylene Glycol, Alumina and CNT nanofluids in CPU cooling: Experimental study", *Experimental Thermal and Fluid Science*, vol. 57, pp. 371–377, 2014. https://doi.org/10.1016/j.expthermflusci.2014.06.003
3. C.J. Ho, L.C. Wei, Z.W. Li, "An experimental investigation of forced convective cooling performance of a microchannel heat sink with Al2O3/water nanofluid", *Applied Thermal Engineering*, vol. 30(2–3), pp. 96–103, 2010. https://doi.org/10.1016/j.applthermaleng.2009.07.003
4. J.F. Tullius, Y. Bayazitoglu, "Effect of Al2O3/H2O nanofluid on MWNT circular fin structures in a minichannel", *International Journal of Heat and Mass Transfer*, vol. 60(1), pp. 523–530, 2013. https://doi.org/10.1016/j.ijheatmasstransfer.2013.01.035
5. S. Vanapalli, H.J.M. Ter Brake, "Assessment of thermal conductivity, viscosity and specific heat of nanofluids in single phase laminar internal forced convection", *International Journal of Heat and Mass Transfer*, vol. 64, pp. 689–693, 2013. https://doi.org/10.1016/j.ijheatmasstransfer.2013.05.033
6. R. Chein, J. Chuang, "Experimental microchannel heat sink performance studies using nanofluids", *International Journal of Thermal Sciences*, vol. 46(1), pp. 57–66, 2007. https://doi.org/10.1016/j.ijthermalsci.2006.03.009
7. R. Chein, G. Huang, "Analysis of microchannel heat sink performance using nanofluids", *Applied Thermal Engineering*, vol. 25(17–18), pp. 3104–3114, 2005. https://doi.org/10.1016/j.applthermaleng.2005.03.008
8. Y. Yu, T. Simon, T. Cui, "A parametric study of heat transfer in an air-cooled heat sink enhanced by actuated plates", *International Journal of Heat and Mass Transfer*, vol. 64, pp. 792–801, 2013. https://doi.org/10.1016/j.ijheatmasstransfer.2013.04.065
9. S.A. Jajja, W. Ali, H.M. Ali, A.M. Ali, "Water cooled minichannel heat sinks for microprocessor cooling: Effect of fin spacing", *Applied Thermal Engineering*, vol. 64(1–2), pp. 76–82, 2014. https://doi.org/10.1016/j.applthermaleng.2013.12.007
10. L. Lin, J. Zhao, G. Lu, X.D. Wang, W.M. Yan, "Heat transfer enhancement in microchannel heat sink by wavy channel with changing wavelength/amplitude", *International Journal of Thermal Sciences*, vol. 118, pp. 423–434, 2017. https://doi.org/10.1016/j.ijthermalsci.2017.05.013

11. K.K. Sikka, K.E. Torrance, C.U. Scholler, P.I. Salanova, "Heat sinks with fluted and wavy plate fins in natural and low-velocity forced convection", *IEEE Transactions on Components and Packaging Technologies*, vol. 25(2), pp. 283–292, 2002. https://doi.org/10.1109/TCAPT.2002.1010019

12. Md. A.R. Junaidi, R. Rao, S.I. Sadaq, M.M. Ansari, "Thermal analysis of splayed pin fin heat sink", *International Journal of Modern Communication Technologies & Research (IJMCTR)*, vol. 2(4), pp. 48–53, 2014. https://www.erpublication.org/ijmctr/published_paper/IJMCTR021417

13. S.B. Chin, J.J. Foo, Y.L. Lai, T.K.K. Yong, "Forced convective heat transfer enhancement with perforated pin fins". *Heat and Mass Transfer*, vol. 49(10), pp. 1447–1458, 2013. https://doi.org/10.1007/s00231-013-1186-z

14. ANSYS FLUENT 13 User's Guide, *Ansys Fluent Theory Guide, ANSYS Inc.*, USA, vol. 15317(November), pp. 724–746, 2013.

Chapter 16

A comparative study of patient-specific bifurcated carotid artery with different viscosity models

Subash Chand Pal, Manish Kumar and Ram Dayal
Malaviya National Institute of Technology Jaipur, Jaipur, India

CONTENTS

16.1 INTRODUCTION: BACKGROUND AND DRIVING FORCES

Many researchers show that the stagnation and highly disturbed flows are key factors to increase the probability of atheroma buildup nearby the vessel bifurcations [1]. Wall shear stress is the key factor in controlling plaque formation and rupture [2]. The development and rupture of atherosclerotic plaque, which causes stroke, are also influenced by flow disruption and wall shear stress (WSS) [3]. In addition to this, in the present world, the most noticeable cause of death is atherosclerosis. The most common cardiovascular disease, as shown in Figure 16.1, is coronary heart disease (CHD), which is caused by atherosclerotic abrasions obstructing the coronary arteries. [4]. These data and findings motivate researchers to develop a technique for estimating WSS of carotid artery in vivo.

It is well accepted that recirculation zone and oscillatory flow-induced WSSs within a blood vessel leads to the development of atherosclerosis [5] Furthermore, at present, the medical community relies on conventional

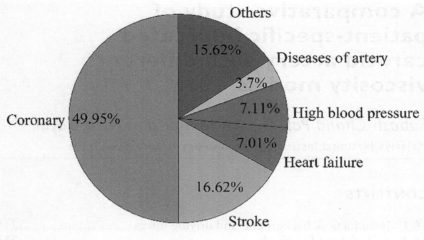

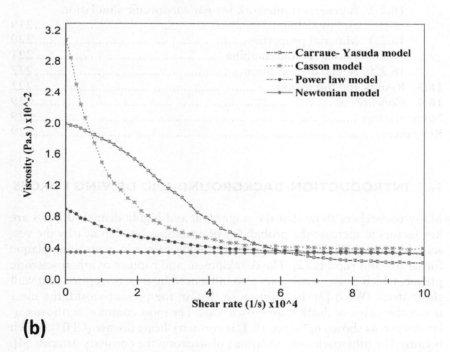

Figure 16.1 (a) Representation of various cardiovascular diseases [4]; (b) relation between blood viscosity and shear rate for different viscosity models.

diagnosis methodology to detect cardiovascular-related diseases (like aortic aneurysm, cardiac arrest, atherosclerosis, and heart murmurs) due to which the diseases are identified at a later stage [6]. To treat, diagnose, and prevent illnesses, it is critical to understand the mechanics of flow and the related WSS. Conventionally, tracing of red blood cells using video monitoring, ultrasound Doppler method, and Doppler velocimetry (LDV) are used in in vivo measurement of the flow velocity [7]. However, the Doppler ultrasound method provides poor spatial resolution when measuring the flow velocity profile. In addition to these drawbacks, WSS cannot be evaluated directly in vivo [8]. In this context, numerical simulation may be used to measure WSS and can be used to anticipate and diagnose cardiovascular disease early [9]. In recent years, numerical simulations have gained a lot of attention as a way to explore WSS because they enable for patient-specific geometry and realistic boundary conditions for blood flow analysis [10]. However, wall elasticity, the use of an adequate viscosity model, realistic boundary conditions, and correct anatomic models are all important factors in the analysis of blood flow in patient-specific models [11].

Three-dimensional CFD simulation over patient-specific models can accurately estimate the flow rate with realistic boundary conditions [12]. Furthermore, several approaches for implementing physiologically realistic boundary conditions have been proposed. One of the most frequent boundary conditions is the resistance boundary condition, which needs any flow parameter and outlet pressure. In resistance boundary circumstances, however, the pressure waves and the resulting flow must be in phase, which contradicts the wave propagation phenomenon [11].

Shear rate–dependent viscosity models are proposed using the Carreau–Yasuda, Power-law, and Casson models. Many investigations revealed that viscosity has a minor nonlinear effect in big arteries like the aorta [8]. Non-Newtonian assumptions have no influence on the shear stress of the walls at average Reynolds numbers for blood flow, according to a study of blood flow through coronary segments [13]. The application of non-Newtonian assumptions was only shown and validated in narrow arteries. An investigation on the bifurcated carotid artery revealed that the Casson model is only required for portraying blood characteristics when the shear rate is less than 10 s^{-1}; when the shear rate is more than this, flow characterization is unaffected, and blood exhibits only Newtonian fluid properties [14]. Therefore, this chapter focused to investigate the nonlinear effect on hemodynamic parameters.

In this work, CFD simulation of patient-specific bifurcation artery with different viscosity models is performed with mentioned realistic boundary conditions. Creation of patient-specific arterial model and the subsequent determination of the flow physics inside it required several steps. It includes clinical imaging, segmentation and reconstruction discretization; boundary conditions; simulation; and post-processing. In this study, 3D patient-specific geometry is generated by segmentation of magnetic resonance imaging scan with an open-source software Simvascular. Then, exported geometry is imported in ANSYS Workbench to perform meshing. First simulation was run with 314580 tetrahedral elements. Mesh convergence was performed to get mesh-independent results. After confirming mesh-independent results, simulations are run with Newtonian, non-Newtonian, Power-law, and Carreau viscosity models.

The major goal of this research is to see how different viscosity models affect hemodynamic parameters in a patient-specific bifurcated artery with realistic border conditions. Then, the comparisons between Newtonian and different non-Newtonian viscosity models are presented based on velocity profile, WSS, and stream lines distributions.

16.2 MATERIALS AND METHODS

16.2.1 Blood flow modeling

For mathematical blood flow modeling, Navier-Stokes equations are utilized. The following are the three-dimensional conservation equations for an incompressible fluid:

Conservation of mass:

$$\Delta \cdot v = 0 \tag{16.1}$$

Conservation of momentum:

$$\rho \left(\frac{\partial v}{\partial t} + v \cdot \nabla v \right) = \nabla \cdot \sigma + \mathrm{f}, \tag{16.2}$$

where the density of the fluid ρ, the viscosity vector v, the stress tensor σ, and f are body force, which are negligible in this simulation. In terms of hydrostatic and deviatoric stress, the stress tensor may now be written as:

$$\sigma = -pI + \tau, \tag{16.3}$$

where the pressure p, the identity tensor I, and the deviatoric stress tensor τ may be written in terms of shear rate tensor (D)

$$\tau = \mu(\dot{\gamma})D, \tag{16.4}$$

where the dynamic viscosity of blood μ, the shear rate $\dot{}$, and shear rate tensor (D) may now be defined as follows:

$$D = \frac{1}{2}(\nabla v + \nabla v^{\mathrm{T}}) \tag{16.5}$$

Finally, $\dot{\gamma}$ is defined in terms of D as:

$$\dot{\gamma} = \sqrt{\frac{1}{2}\sum_i\sum_j D_{ij}D_{ji}} \qquad (16.6)$$

The viscosity of blood can be calculated with the help of consecutive equations [14]. The Newtonian fluid model, which assumes a constant viscosity of blood, is the simplest, whereas the Casson models, power-law model, and Carreau–Yasuda models are employed for non-Newtonian blood. The power-law model may be written as follows:

$$\mu = k\dot{\gamma}^{n-1} \qquad (16.7)$$

where the power-law index, n, and the flow consistency index, k, represent non-Newtonian behavior of blood. The Casson model simulates blood flow in a small artery at low shear rates by taking into consideration blood's shear thinning behavior and yield stress. The following equation can be used to express this model.

$$\mu = \left\{\sqrt{\frac{\tau_0}{I\dot{\gamma}I}[1 - e^{(-m\dot{\gamma})}]} + \sqrt{\mu_0}\right\}^2 \qquad (16.8)$$

where μ_0 is the Newtonian viscosity, τ_0 is the yield stress, and when shear rate approaches zero, m determines viscosity. The Carreau–Yasuda model is an extension of Newtonian law and looks like a power-law. The experimental results on the viscosity and shear rate connection best suited the Carreau–Yasuda model among these four models [8].

$$\frac{\mu - \mu_\infty}{\mu_0 - \mu} = (1 + (\lambda\dot{\gamma})^a)^{n-\frac{1}{a}} \qquad (16.9)$$

Table 16.1 summarizes the parameters utilized in each viscosity model. In addition, all viscosity models are demonstrated in Figure 16.1.

16.2.2 A general framework for patient-specific simulation and modeling

In this study, patient-specific segmentation and modeling were created using open-source software Simvascular in various steps, as shown in Figure 16.2. DICOM file is used as input and then the required path is created for segmentation to create an appropriate anatomical model. To get a smoother model, the next step is to iteratively run a number of smoothing processes. Created model is exported as.STL file and meshing is performed in ANSYS

Table 16.1 Details of different viscosity model parameters [14]

Viscosity model	Parameters
Newtonian model	$\mu = 0.0035\,\text{Pa.s}$
Casson model	$\tau_0 = 0.004\,\text{Pa}, \mu_0 = 0.004\,\text{Pa.s}, m = 100$
Power-law model	$k = 0.0035, n = 0.6$
Carreau–Yasuda model	$\mu_\infty = 0.002\,\text{Pa s}\ \ \mu_0 = 0.022\,\text{Pa.s}$
	$\lambda = 0.1\,\text{ls}, n = 0.392, a = 2$

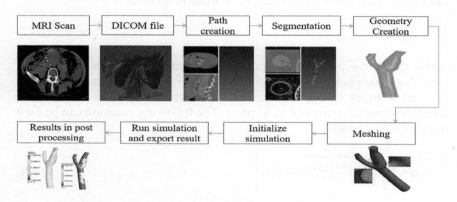

Figure 16.2 General framework for patient-specific simulation and modeling.

Workbench. The investigation of viscosity models is another important factor. Impact of Newtonian and different non-Newtonian viscosity models have been assessed in this chapter, and comparison of the velocity profile and WSS is shown in the Results and Discussion section. Furthermore, it is recommended that the boundary layer should be fully developed before it reaches the point of bifurcation in the geometry. This is so because, if the length is not sufficient, actual hemodynamic behavior inside the artery cannot be represented accurately. Therefore, sufficient length of aorta is considered to support the development of the boundary layer.

16.2.3 Material properties

In order to explore the effect of Newtonian and non-Newtonian viscosity models on a blood flow through bifurcated artery, different simulations were performed.

Blood was modeled as Newtonian fluid with density, $\rho = 1060$ kg/m³, dynamic viscosity $\mu = 0.0035$ Pa·s. It is modeled as a non-Newtonian fluid with non-Newtonian power-law (density $\rho = 1060$ kg/m³, consistency index = 0.0167, power-law index $n = 0.6$, and minimum viscosity $\mu_{\min} = 0.0035$

Pa·s and maximum viscosity μ_{max} = 0.056 Pa·s) and with Carreau model (density ρ = 1060 kg/m³, time constant lambda = 0.03568, zero shear viscosity μ_0 = 0.0035 Pa·s, and infinite shear viscosity μ_∞ = 0.056 Pa·s).

16.2.4 Geometry and meshing

The geometry of bifurcation artery is shown in Figure 16.3. It is created with the help of open-source software Simvascular; the detailed process is shown in Figure 16.2. Finally, geometry is imported in ANSYS Workbench for meshing and simulation. The inlet is represented by the CCA (common carotid artery). The ECA (external carotid artery) and ICA (internal carotid artery) are the bifurcation's downstream outputs of carotid artery. In the ICA, which is shown in Figure 16.2, there is a carotid sinus. The geometry's dimensions are shown in Table 16.2.

The fluid domain is discretized into 11,84,205 tetrahedral elements for CFD analysis. The grid-independent analysis was carried out till grid

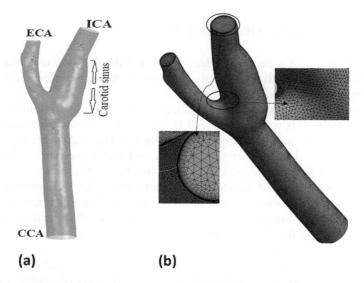

(a)   **(b)**

Figure 16.3 Carotid bifurcated artery. (a) Geometry; (b) Meshing.

Table 16.2 Sectional dimensions of the inlet and outlets [15]

Sections	Diameter (mm)	Area (mm²)
CCA	6.27	30.86
ICA	4.30	14.51
ECA	3.10	7.54

independence was attained. Figure 16.4 represents a summary of the convergence study which was performed in the fluid domain based on average velocity at the outlet of ICA and ECA. This parameter is chosen due to interest of this study.

The results reveal that increasing the fluid mesh elements from 11,84,205 to 15,24,876 has just a 0.457% and 0.267% influence on the velocity magnitude at the outlet of ICA and ECA. Furthermore, increasing the number of mesh elements required longer to run the model, and the variations in velocity results are extremely little, which does not justify the processing expense.

16.2.5 Boundary condition

The cardiac cycle is repeated every 1 s at a heart rate of 60 beats per minute. The pulsatile nature of the human heart cycle was applied at the inlet, CCA, with two distinct phases of diastole and systole at $t = 0.21$ s indicating maximum flow rate illustrated in Figure 16.4 [15]. For the outlet ICA and ECA, open boundary condition applied with zero-gauge pressure at outlet.

16.3 RESULTS AND DISCUSSION

Over the period of a cardiac cycle, Figure 16.5 displays the distribution of mass flow rate through the outputs ECA and ICA for Newtonian and non-Newtonian viscosity models.

For validation of results, pulsatile nature of the Newtonian fluid is used to compare the results with reference [15], Then, for each non-Newtonian viscosity model, a separate analysis is performed. The ICA (internal carotid artery) receives the majority of the flow, and the highest mass flow rate occurs during the systolic peak ($t = 0.21$ s) of the cardiac cycle.

It can be seen that the Newtonian viscosity model shows maximum mass flow rate at any instance of cardiac cycle, and the Carreau viscosity model represents a minimum mass flow rate which indicates that the fluid behaves as shear thinning for the Carreau model.

Figure 16.6 depicts the magnitude of velocity during the cardiac cycle at three separate locations in the fluid domains of CCA, ICA, and ECA. Among all of these sites, the velocity magnitude in the ICA at systolic point is shown to be considerably greater than in the CCA and ECA at systolic point.

Throughout the cardiac cycle, no significant differences in velocity magnitude were detected for different viscosity models.

The axial velocity vectors are plotted on the different slices and presented in Figure 16.7, which represents the fully developed parabolic nature of velocity profile at the inlet, with a maximum velocity at the section's center and progressively decreasing to zero at the wall. A relatively low velocity is found near the bifurcation in the ICA, indicating that blood is recirculating in this region. In Figure 16.8, the recirculation zone can also be observed from the

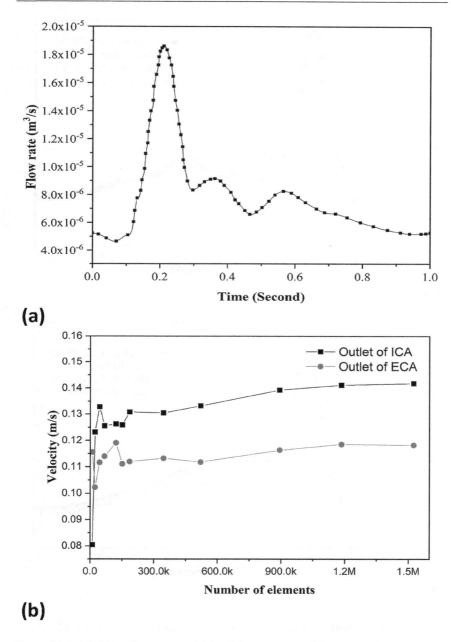

Figure 16.4 (a) Pulsatile nature of blood flow rate profile imposed at the inlet, CCA; (b) variation of velocity magnitude with mesh refinement.

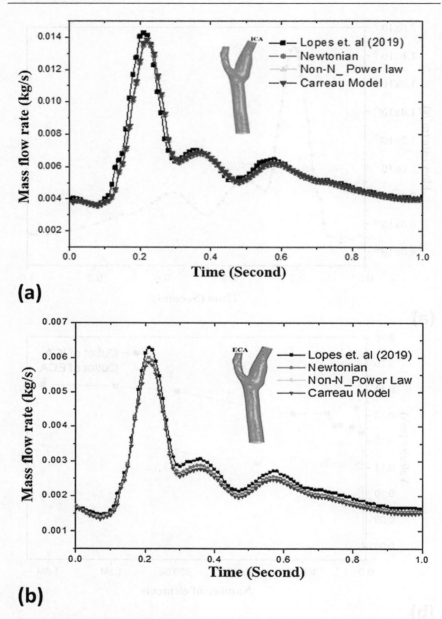

Figure 16.5 Distribution of mass flow rate to the outlets, (a) ICA; (b) ECA.

stream line at the systolic peak. It represents that stream lines are recirculating at the carotid sinus where velocity is lower and artery is expanded.

Figure 16.9 depicts the WSS at $t = 0.21$ s, which corresponds to the systolic peak. It illustrates that low WSS exists near the bifurcation in the ICA

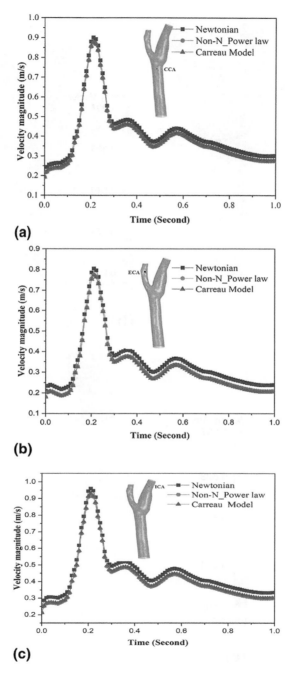

Figure 16.6 Velocity magnitude presented at different points within the fliud domain (a) CCA; (b) ECA; (c) ICA.

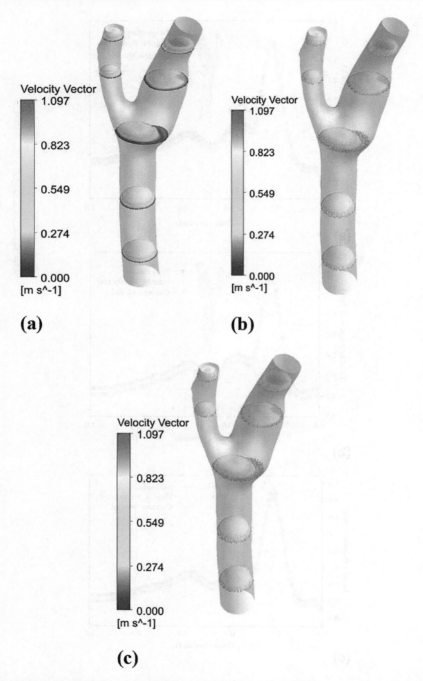

Figure 16.7 Velocity vector at different planes at systolic peak. (a) Newtonian model; (b) non-Newtonian power-law model; (c) Carreau–Yasuda model.

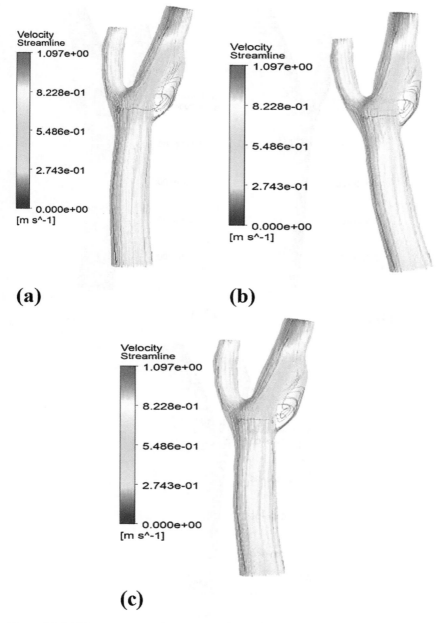

Figure 16.8 Velocity stream line at systolic peak. (a) Newtonian model; (b) non-Newtonian power-law model; (c) Carreau–Yasuda model.

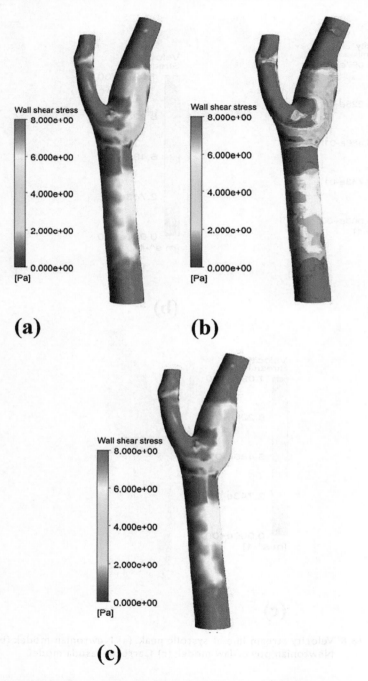

Figure 16.9 Wall shear stress profile at systolic peak. (a) Newtonian model; (b) non-Newtonian power-law model; (c) Carreau–Yasuda model.

where recirculation of blood exists, as shown in Figure 16.8, and it leads to the formation of plaque.

16.4 CONCLUSIONS

A comparative study of Newtonian and different non-Newtonian viscosity models for patient-specific bifurcated carotid arteries is presented in this work. The following are the major findings of this study:

- The carotid sinus with reduced WSS and blood recirculation is more prone to accumulate plaque and develop atherosclerosis.
- Velocity magnitude at the center of artery with Newtonian viscosity model is higher than that of non-Newtonian viscosity models.
- The magnitude of velocity in the ICA at systolic point is observed to be substantially higher than in the CCA and ECA at systolic point.

NOMENCLATURE

t	Time	(s)
ρ	Density of blood	[kg/m^3]
μ	Dynamic viscosity of blood	[Pa·s]
n	Power-law index	–
ICA	Internal carotid artery	–
CCA	Common carotid artery	–
ECA	External carotid artery	–

REFERENCES

1. P. Papathanasopoulou, S. Zhao, U. Kohler, et al., MRI measurement of time-resolved wall shear stress vectors in a carotid bifurcation model, and comparison with CFD predictions. *Journal of Magnetic Resonance Imaging* 17(2), 2003, 153–162.
2. D.P. Giddens, C.K. Zarins, S. Glagov, The role of fluid mechanics in the localization and detection of atherosclerosis. *Journal of Biomechanical Engineering* 115, 1993, 588–594.
3. D. Bluestein, Y. Alemu, I. Avrahami, M. Gharib, K. Dumont, J.J. Ricotta, S. Einav, Influence of micro calcifications on vulnerable plaque mechanics using FSI modeling. *Journal of Biomechanics* 41, 2008, 1111–1118.
4. American Heart Association, Heart disease and stroke statistics. *Circulation* 123(4), 2011, 18–209.

5. L. Sousa, C. Castro, C. Antonio, R. Chaves, Computational techniques and validation of blood flow simulation. *WSEAS Transactions on Biology and Biomedicine* 8(4), 2011, 145–155.

6. V. Kanyanta, N. Quinn, S. Kelly, et al, Fluid-Structure Interaction (FSI) in bioengineering. *Procedia Structural Integrity* 24, 2014, 939–948.

7. S.J. Lee, H.J. Ha, In vivo measurement of blood flow in a micro-scale stenosis model generated by laser photothermal blood coagulation. *IET Systems Biology* 7 (2), 2012, 50–55.

8. S.K. Shanmugavelasyudam, D. Rubenstein, W. Yin, Effect of geometrical assumptions on numerical modelling of coronary blood flow under normal and disease conditions. *Journal of Biomechanical Engineering* 132, 2010, 1–8.

9. V. Hollander, D. Durban, et al., Experimentally validated microstructural 3D constitutive model of coronary arterial media. *Journal of Biomechanical Engineering* 133(3), 2011, 1–14.

10. J. Dong, K.K.L. Wong, J. Tu, Hemodynamics analysis of patientspecifc carotid bifurcation A CFD model of downstream peripheral vascular impedance, *International Journal* for *Numerical Methods* in *Biomedical Engineering* 29, 2013, 476–491.

11. V. Clementel, C.A. Figueroa, K.E. Jansen, C.A. Taylor, Outflow boundary conditions for 3D simulations of nonperiodic blood flow and pressure fields in deformable arteries. *Computer Methods in Biomechanics and Biomedical Engineering* 13, 2010, 625–640.

12. C.A. Taylor, M.T. Draney, Experimental and computational methods in cardiovascular fluid mechanics. *Annual Review of Fluid Mechanics* 36, 2004, 197–231.

13. A.K. Chaniotis, L. Kaiktsis, D. Katritsis, E. Efstathopoulos, Computational study of pulsatile blood flow in prototype vessel geometries of coronary segments. *An International Journal Devoted to the Applications of Physics to Medicine and Biology* 26(3), 2010, 140–156.

14. H. Gharahi, B.A. Zambrano, D.C. Zhau, Computational fluid dynamic simulation of human carotid artery bifurcation based on anatomy and volumetric blood flow rate measured with magnetic resonance imaging. *International Journal of Advances in Engineering Sciences and Applied Mathematics*, DOI 10.1007/s12572-016-0161-6, 2016.

15. D. Lopes, H. Pugas, J.C. Teixeira, S.F. Teixeira, Influence of arterial mechanical properties on carotid blood flow: Comparison of CFD and FSI studies. *International Journal of Mechanical Sciences* 160, 2019, 209–218.

Chapter 17

Recent advancements in induced-charge electrokinetic micromixing

Anshul Kumar Bansal, Ram Dayal and Manish Kumar

Malaviya National Institute of Technology Jaipur, Rajasthan, India

CONTENTS

17.1 INTRODUCTION: BACKGROUND AND APPLICATIONS

A homogenous and rapid mixing of fluid in a microchannel is crucial for many biomedical and chemical applications such as PCR amplification, DNA hybridization, drug discovery, analysis of disease symptoms, cell lysis, biological screening, enzyme assays, and protein folding [1, 2]. Today, lab on a chip and micro-total analysis system (μ-TAS) technologies are widely used for miniaturization in biomedical and biochemical analysis. It has many advantages like less sample and reagent consumption, accurate and fast results, reproducibility, and reliability. Flows in microfluidic devices are an order of low Reynolds number laminar flow; therefore, mixing in microfluidic devices is related to molecular diffusion, which is a prolonged process and requires a long channel [3]. Different types of techniques are used for rapid mixing in the microchannel, which is categorized into passive and active micromixers (Figure 17.1).

Passive micromixer does not require any external power source and mixes the fluids with different methods like microchannel with complex geometry [4], parallel and serial injection [5], grooves in microchannel wall [6], and flow recombination [7]. These methods increase the interferential contact within the mixing fluid streams and enhance the mixing efficiency. Although passive micromixers have easy design and simple implementation,

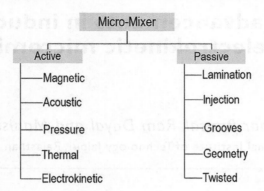

Figure 17.1 Classification of micromixer.

microchannel increases the mixing performance but still requires long channel and sufficient time to mix the fluids. Active micromixer requires external power source using magnetic field [8], pressure disturbance [9], ultrasonic (acoustic disturbance) [10], thermal disturbance [11], and electrokinetic [12]. Active methods provide a better mixing index in small-length microchannel within a short time. Most of the active micromixing methods raise the temperature of mixing samples, which is an adverse effect on the sample properties, hence less suitable for biological applications. Most of the above active micromixers require external variable drivers, which are difficult to integrate into the microfluidic system. In active micromixer, electrokinetic micromixer has shown great advantages over other methods. Ease of control, no moving component, facile integration with the microfluidic system, no mechanical system (no fatigue and vibration), and no sample dispersion are the main benefits of electrokinetic micromixer.

Electrokinetic micromixers use different methods to improve the mixing performance, such as geometrical modification, heterogeneous wall surface zeta-potential, and electrokinetic instability using a time-dependent electric field. Chen and Cho [13] find out the mixing performance in a different configuration of the wavy surface of the microchannel with electrokinetically driven flow. Mixing performance is improved by applying the heterogeneous zeta-potential on the wall and increasing the amplitude and length of a wavy surface portion. Shamloo et al. [14] placed the electrode array at a specific location in a microchannel supplied with alternating current (electrokinetic instability) to improve the mixing index. They analyzed the mixing efficiency for a different geometric feature by using one ring type, double ring type, and diamond shape micromixer chamber with Newtonian and non-Newtonian fluid. They enhance the mixing efficiency by about 99% by optimizing the different parameters such as voltage amplitude and frequency of AC source. Bhattacharyya and Bera [15] investigated the effect of rectangular obstacle and non-uniform surface zeta-potential on mixing index driven by combined pressure and electroosmosis flow. The development of vortical flow near the rectangular block of different zeta-potential enhanced the mixing performance. All of the above electrokinetic methods require complex surface heterogeneity, surface modification, and high voltage control. Induced-charge electrokinetic is another innovative method to use for enhancing mixing performance. When a highly polarizable conductive surface is placed in an external electrical field, it induces the vortex around the conductive surface, improving the mixing in the microchannel.

Electrokinetic micromixers use induce charge phenomenon to play a significant role in the mixing technologies. Many researchers have investigated this phenomenon and given many innovative designs of micromixers to enhance the mixing performance. However, currently, no specific review paper is focused on this area. In the present work, recent development and advancement of the induced-charge electrokinetic micromixer techniques with different design parameters and mixing techniques are comprehensively reviewed in terms of their mixing index, mixing time, and mixing length.

17.2 THEORY AND FUNDAMENTAL

Induced-charge electrokinetic contains the phenomenon of induced micro circulating vortex nearby the conductive surface under the influence of the external electric potential. When a highly polarizable and conductive surface is placed in the electrolyte solution, the electric current derives charged ion from the electrolyte solution to the conductive surface perpendicular to the object. Equal amounts of positive and negative charge ions are derived into the thin layer adjacent to the conductive surface by the electric current and generate a dipolar ions layer attached to the conductive surface, as shown in Figure 17.2(a). Because of this dipolar layer, electric field lines are repelled from the charged dipolar object and therefore electric current into the conductive surface is reduced. At steady-state conditions, the conductor is fully polarized and acts as an insulator because the induced screening cloud be developed (Figure 17.2(b)). The induced-charge electric field is equal and opposite to the applied external field ($\vec{E}_{induced} = -\vec{E}_{applied}$) and the induced zeta-potential is varied with the location on a conductive surface. The induced field generated by applied external potential is given by [17].

$$\nabla \zeta_{\text{induced}} = -\nabla \phi_{\text{e}} \tag{17.1}$$

where ζ_{induced} is the induced zeta-potential and ϕ_e is external applied electric potential. Conductive surface is initially electric neutrality; therefore, integration of induced zeta-potential on whole conductor surface S should be zero [17]:

$$\int_S \zeta_{\text{induced}} dS = 0 \tag{17.2}$$

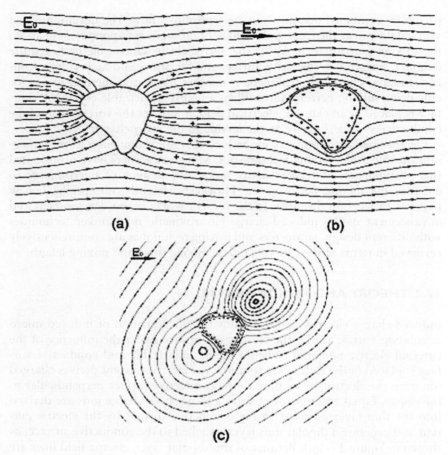

Figure 17.2 ICEK phenomenon: (a) conductive surface placed in electric field; (b) steady state; (c) vortices generated near conductive object [16].

Induced zeta-potential distribution ζ_{induced} is given by [17]:

$$\zeta_{\text{induced}} = -\phi_e + \phi_c \tag{17.3}$$

where ϕ_c is constant and evaluated from Equation (1) and Equation (3)

$$\phi_c = \frac{\int_s \phi_e dA}{A} \tag{17.4}$$

Induced zeta-potential on the conductive object is non-uniform. When external electric field is applied, non-uniform slip velocity is generated, which causes the induced micro-vortices around the conductive surface (Figure 17.2(c)). The slip velocity $\vec{U}_i$ is given by [18].

$$\vec{U}_i = -\frac{\varepsilon\varepsilon_0\zeta_{\text{induced}}}{\mu}\vec{E}_{applied} \qquad (17.5)$$

where ε is dielectric constant, ε_0 is the vacuum permittivity, μ is viscosity. The generated induced vortices enhance the interferential contact of mixing samples and enhance mixing characteristics. The mixing index (ε) for mixing quality in the microchannel is given by [19]:

$$\varepsilon = \left(\frac{1 - \int_0^W (\vec{C} - \vec{C}_\infty)}{\int_0^W (\vec{C}_0 - \vec{C}_\infty)} \right) \times 100\%.$$

where concentration at two inlets A and B are $\vec{C_A} = C_0$ and $\vec{C_B} = 0$, respectively, $\vec{C}_\infty = (\vec{C_A} + \vec{C_B})/2$ and $\vec{C}$ is local concentration at the outlet section.

17.3 ADVANCEMENT IN ICEK

Many researchers have studied the effect of the induced charge electrokinetics (ICEK) phenomenon in micromixing. Induced micro-vortices near the polarized surface increase the chaotic movement in the microchannel and improve mixing. Induced-charge electrokinetic has shown great potential in micromixing because of its easy fabrication, control, and flexibility. Induced-charge micromixer is reviewed in two categories: mixing with conductive chamber and conductive link. Induced-charge micromixer is examined in two categories: mixing with conductive chamber and conductive link.

17.3.1 ICEK mixing with conductive chamber

Many researchers have investigated the conductive chamber and conductive hurdles effect on the mixing performance in the microfluidic device. Wu and Li [16] determined the mixing index for a micromixer having embedded conductive hurdle of different shapes in a rectangular microchannel. The conductive hurdle generated the vortices near the conductive hurdle as shown in Figure 17.3, which enhances the mixing by recirculating the flow. Experimental setup validates the mixing results. Mixing with triangular, rectangular, and circular shaped conductive hurdle is investigated and found that the rectangular hurdle gives an optimum 94.2% mixing index for triple hurdle at 150 V/cm electric fields.

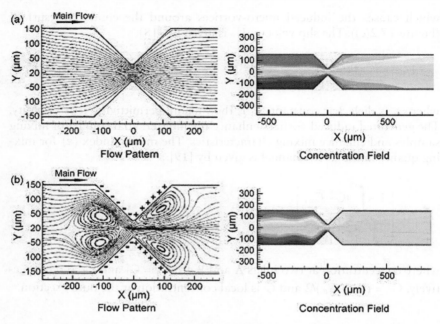

Figure 17.3 Streamline near (a) non-conductive surface and (b) conductive surface [16].

Nazari et al. [20] investigated the effect of the microchannel with circular, square., triangular, and rhomboid conductive chambers with or without conductive edges on the mixing. Mixing performance for different conductive edges and chamber size is determined. It is reported that this type of micromixer has a simple design fabrication, and optimum mixing efficiency of 95% is achieved with a rhomboid type mixing chamber with a single conductive edge. Shamloo et al. [21] designed a 3D T-shaped micromixer with a conduction hurdle at T-junction in a microchannel. Results for different parameters such as inlet angle, aspect ratio, shape, and height are analyzed. Vortices generated near the conductive hurdle enhance the mixing (Figure 17.4). It is reported that optimum mixing quality is achieved for circular hurdle, inlet angle of 180°, and triangular chamber of mixing index of 81.49%.

Jain et al. [22] investigated the mixing enhancement in micromixers with conductive hurdle with different shapes. Vortices are produced near the conductive hurdles resulting in better mixing due to an increase in the interferential contact between fluids. The mixing efficiency of 98% is reported for the rectangular-shaped conductible obstacle. Wu and Li [23] examined a rectangular micromixer with different non-symmetric triangle-shaped electrically conductive hurdle and dimensions of the convergence and divergence shape. Non-uniform slip velocity on conductive hurdle induces the vortex in the flow field and enhances the mixing.

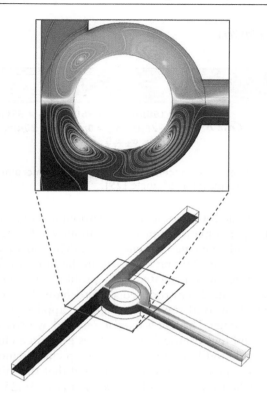

Figure 17.4 Vortices near conductive chamber section [21].

17.3.2 ICEK mixing with conductive link

Many researchers use conductive flap/link fixed or movable in micromixer to enhance the mixing. Kazemi et al. [24] have investigated the effect of electrically conductive flap mounted in the Y-shaped mixing chamber on mixing performance. The impact of the flap position, installation, and electric current intensity is analyzed, and the optimum mixing index is found. It is reported that placing a flap near the throat region and generating micro-vortices improved mixing performance. Nazari et al. [25] installed an electrically conducting plate in a microchannel to enhance the mixing index. Micro-vortices are induced near the conductive plate, making mixing fast and in a small channel length. The parameter including mounting of the plate, length, position, and the number of plates is considered, and the optimal value of mixing index of 99.6% was found by mounting two plates at a 5° angle near the wall region and one more plate at the middle of the microchannel (Figure 17.5).

Najjaran et al. [26] have examined a micromixer with corrugated wall rectangular, triangular, sinusoidal, and trapezoidal shape along with conductive plate under applied electric field. The conductive plate inside the microchannel generates the micro-vortices and enhances the mixing

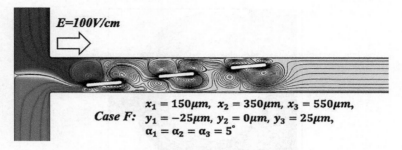

$$x_1 = 150\mu m, \quad x_2 = 350\mu m, \quad x_3 = 550\mu m,$$
Case F: $\quad y_1 = -25\mu m, \quad y_2 = 0\mu m, \quad y_3 = 25\mu m,$
$$\alpha_1 = \alpha_2 = \alpha_3 = 5^\circ$$

Figure 17.5 Streamlines and concentration in T-shaped micromixer with three conductive plates at 5° angle [25].

performance by about 200% compared to without installing the conductive plate. The mixing performance improved by about 9% by increasing potential 20–40 V/cm and reducing efficiency by 0.38% and by decreasing the electric field 40–50 V/cm. Mixing efficiency in the range of 85.45–97.53% is achieved for different corrugated wall profiles with conductive plate installation. Azimi et al. [27] introduced a microchannel with a flexible conductive link, which swayed the area under the applied variable external time-varying DC voltage. The link is fixed at one end, which can move under the electric field and hydrostatic force. Vortices generated close to the conductible flexible link and link movement develop chaotic movement in the flow to improve the mixing index. It is reported that the mixing efficiency improved up to 90% for a conductible link length of 156.25 μm.

Goodarzi et al. [26] used a curved conductive plate inside the T-shaped micromixer. The curve surface radius, span length, pattern arrangement, and orientation angle relative to the flow significantly affect the mixing performance. The mixing efficiency for the three curved conductive surfaces with the same concavity direction is 91.86%. In contrast, the flipped one surface concavity of the middle curve plate increased the mixing efficiency to 95.44%. Daghighi and Li [19] proposed a micromixer with an electrically conductive particle inside a mixing chamber. Micro-vortices are created on the electrically conductible surface, as shown in Figure 17.6. It creates perturbation in the flow and enhances the mixing performance. The applied electric field at a 45° angle reduces the mixing time.

17.4 CONCLUSION

This chapter summarized the available literature on induced-charge electrokinetic micromixers (Table 17.1).

Some critical remarks are given below:

- In contact with an electrolyte solution, surfaces get charged due to ion absorption from electrolyte or ionization of surface group. Counter

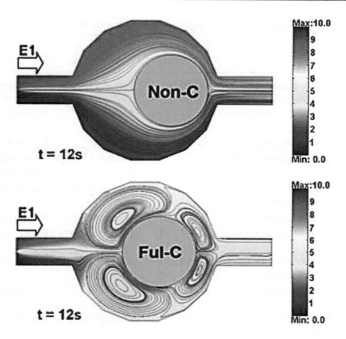

Figure 17.6 Flow field around non-conductive and conductive particle [19].

ions from electrolytes make surface charge neutralized and form an electrical double layer. By applied electric field, EDL free ions make fluid flow (electroosmosis flow). Circulation or vortices are generated due to different surface charges on the wall, enhancing the mixing performance.

- Installing conductive link or particle inside the micromixer induced the micro-vortices around it in the fluid stream; hence, chaotic movement increases the mixing tendency. It makes micromixing fast and in a short length of microchannel.
- Length, orientation, and the number of conductive links also significantly affect the mixing performance. The applied electrical field also increases and decreases mixing efficiency in term of their range.
- Conductive chamber or obstacle with different shapes increase the vortices generation inside the fluid flow and creates disturbances inside the fluid flow. Therefore, it enhanced the effectiveness of the electrokinetic micromixer.

Following important points are suggested for future research:

- Most of the studies in micromixing are commonly focused on Newtonian fluids, while all biological fluids are non-Newtonian. Hence, more attention is required for non-Newtonian mixing.

Table 17.1 Micromixing in induced-charge electrokinetic micromixer

Characteristics	Mixing efficiency	References	Setup
Rectangular microchannel with embedded conducting hurdles	94.2%	Wu and Li [16]	Numerical and experimental
Mixing chamber with conductive surfaces	95.1%	Nazari [20]	Numerical
Obstacle shape optimization	98%	Jain [22]	Numerical
Pair of conducting triangle hurdles	–	Wu and Li [23]	Numerical
T-shaped with conductive hurdle	81.49%	Shamloo [21]	Numerical
Y-shape micromixer with electrically conductive flap	–	Kazemi [24]	Numerical
T-shaped with conductive plate	99.6%	Nazari [25]	Numerical
Corrugated walls with conductive plate	97.53%	Najjaran [27]	Numerical
Continuous deformation of a conducting flexible link	93.24%	Azimi [28]	Numerical
T-micromixer with conductive curved arc plate	95.44%	Goodarzi [26]	Numerical
Micromixer with an electrically conducting particle	>95%	Daghighi [19]	Numerical
Conductive triangular-posts in a straight channel	64%	Harnett [29]	Numerical and experimental
Non-uniform electric field on a conducting cylinder	96.89%	Manshadi [30]	Numerical

- Non-uniform channel profile and conductive surface location significantly affect the mixing efficiency. Therefore, optimal design of these parameters is necessary for efficient micromixing.
- All biological fluids are sensitive to temperature; therefore, heat generation and thermal aspects are necessary.
- There are several passive techniques to increase the mixing index. Combination with induced-charge electrokinetic active techniques increases the mixing significantly.
- In most literature, mixing time is not mentioned, whereas biological reactions depend on time. Therefore, mixing time should also be the main criterion for designing a micromixer.

REFERENCES

1. C. Y. Lee and L. M. Fu, "Recent advances and applications of micromixers," *Sensors Actuators, B Chem.*, vol. 259, pp. 677–702, 2018, doi: 10.1016/j.snb.2017.12.034.

2. G. S. Jeong, S. Chung, C.-B. Kim, and S.-H. Lee, "Applications of micromixing technology," *Analyst*, vol. 135, no. 3, pp. 460–473, Feb. 2010, doi: 10.1039/ B921430E.

3. M. Hadigol, R. Nosrati, A. Nourbakhsh, and M. Raisee, "Numerical study of electroosmotic micromixing of non-Newtonian fluids," *J. Nonnewton. Fluid Mech.*, vol. 166, no. 17–18, pp. 965–971, 2011, doi: 10.1016/j.jnnfm.2011.05.001.

4. N. Aoki and K. Mae, "Effects of channel geometry on mixing performance of micromixers using collision of fluid segments," *Chem. Eng. J.*, vol. 118, no. 3, pp. 189–197, 2006, doi: 10.1016/j.cej.2006.02.011.

5. J. J. Chen and Y. S. Shie, "Interfacial configurations and mixing performances of fluids in staggered curved-channel micromixers," *Microsyst. Technol.*, vol. 18, no. 11, pp. 1823–1833, 2012, doi: 10.1007/s00542-012-1489-x.

6. C. A. Cortes-Quiroz, A. Azarbadegan, M. Zangeneh, and A. Goto, "Analysis and multi-criteria design optimization of geometric characteristics of grooved micromixer," *Chem. Eng. J.*, vol. 160, no. 3, pp. 852–864, Jun. 2010, doi: 10.1016/J.CEJ.2010.02.029.

7. X. Chen, T. Li, H. Zeng, Z. Hu, and B. Fu, "Numerical and experimental investigation on micromixers with serpentine microchannels," *Int. J. Heat Mass Transf.*, vol. 98, pp. 131–140, Jul. 2016, doi: 10.1016/J. IJHEATMASSTRANSFER.2016.03.041.

8. L. H. Lu, K. S. Ryu, and C. Liu, "A magnetic microstirrer and array for microfluidic mixing," *J. Microelectromech. Syst.*, vol. 11, no. 5, pp. 462–469, 2002, doi: 10.1109/JMEMS.2002.802899.

9. Y. Ma et al., "An unsteady microfluidic T-form mixer perturbed by hydrodynamic pressure," *J. Micromech. Microeng.*, vol. 18, no. 4, 2008, doi: 10.1088/0960-1317/18/4/045015.

10. G. G. Yaralioglu, I. O. Wygant, T. C. Marentis, and B. T. Khuri-Yakub, "Ultrasonic mixing in microfluidic channels using integrated transducers," *Anal. Chem.*, vol. 76, no. 13, pp. 3694–3698, 2004, doi: 10.1021/ac035220k.

11. J. H. Tsai and L. Lin, "Active microfluidic mixer and gas bubble filter driven by thermal bubble micropump," *Sensors Actuators, A Phys.*, vol. 97–98, pp. 665–671, 2002, doi: 10.1016/S0924-4247(02)00031-6.

12. N. Loucaides, A. Ramos, and G. E. Georghiou, "Configurable AC electroosmotic pumping and mixing," *Microelectron. Eng.*, vol. 90, pp. 47–50, 2012, doi: 10.1016/j.mee.2011.04.007.

13. C. C. Cho, C. L. Chen, and C. K. Chen, "Electrokinetically-driven non-Newtonian fluid flow in rough microchannel with complex-wavy surface," *J. Nonnewton. Fluid Mech.*, vol. 173–174, pp. 13–20, 2012, doi: 10.1016/j. jnnfm.2012.01.012.

14. A. Shamloo, M. Mirzakhanloo, and M. R. Dabirzadeh, "Numerical Simulation for efficient mixing of Newtonian and non-Newtonian fluids in an electroosmotic micro-mixer," *Chem. Eng. Process. Process Intensif.*, vol. 107, pp. 11–20, 2016, doi: 10.1016/j.cep.2016.06.003.

15. S. Bhattacharyya and S. Bera, "Combined electroosmosis-pressure driven flow and mixing in a microchannel with surface heterogeneity," *Appl. Math. Model.*, vol. 39, no. 15, pp. 4337–4350, 2015, doi: 10.1016/j.apm.2014.12.050.

16. Z. Wu and D. Li, "Micromixing using induced-charge electrokinetic flow," *Electrochim. Acta*, vol. 53, no. 19, pp. 5827–5835, 2008, doi: 10.1016/j. electacta.2008.03.039.

17. Y. Daghighi, Y. Gao, and D. Li, "3D numerical study of induced-charge electrokinetic motion of heterogeneous particle in a microchannel," *Electrochim. Acta*, vol. 56, no. 11, pp. 4254–4262, 2011, doi: 10.1016/j.electacta.2011.01.083.

18. M. A. C. Stuart et al., *Fundamentals of interface and colloid science: Soft Colloids*, vol. 128, no. August 2008, 2010. Published: 29th August 2008 http://bit.ly/1WkK9Av.

19. Y. Daghighi and D. Li, "Numerical study of a novel induced-charge electrokinetic micro-mixer," *Anal. Chim. Acta*, vol. 763, pp. 28–37, 2013, doi: 10.1016/j.aca.2012.12.010.

20. M. Nazari, S. Rashidi, and J. A. Esfahani, "Mixing process and mass transfer in a novel design of induced-charge electrokinetic micromixer with a conductive mixing-chamber," *Int. Commun. Heat Mass Transf.*, vol. 108, no. August, p. 104293, 2019, doi: 10.1016/j.icheatmasstransfer.2019.104293.

21. A. Shamloo, M. Madadelahi, and S. Abdorahimzadeh, "Three-dimensional numerical simulation of a novel electroosmotic micromixer," *Chem. Eng. Process. Process Intensif.*, vol. 119, no. May, pp. 25–33, 2017, doi: 10.1016/j.cep.2017.05.005.

22. M. Jain, A. Yeung, and K. Nandakumar, "Induced charge electro osmotic mixer: Obstacle shape optimization," *Biomicrofluidics*, vol. 3, no. 2, 2009, doi: 10.1063/1.3167279.

23. Z. Wu and D. Li, "Mixing and flow regulating by induced-charge electrokinetic flow in a microchannel with a pair of conducting triangle hurdles," *Microfluid. Nanofluidics*, vol. 5, no. 1, pp. 65–76, 2008, doi: 10.1007/s10404-007-0227-7.

24. Z. Kazemi, S. Rashidi, and J. A. Esfahani, "Effect of flap installation on improving the homogeneity of the mixture in an induced-charge electrokinetic micro-mixer," *Chem. Eng. Process. Process Intensif.*, vol. 121, no. August, pp. 188–197, 2017, doi: 10.1016/j.cep.2017.08.015.

25. M. Nazari, P. Y. A. Chuang, J. Abolfazli Esfahani, and S. Rashidi, "A comprehensive geometrical study on an induced-charge electrokinetic micromixer equipped with electrically conductive plates," *Int. J. Heat Mass Transf.*, vol. 146, p. 118892, 2020, doi: 10.1016/j.ijheatmasstransfer.2019.118892.

26. V. Goodarzi, S. H. Jafarbeygi, R. A. Taheri, M. Sheremet, and M. Ghalambaz, "Numerical investigation of mixing by induced electrokinetic flow in T-micromixer with conductive curved arc plate," *Symmetry*, vol. 13, no. 6, 915, May 2021, doi: 10.3390/SYM13060915.

27. S. Najjaran, S. Rashidi, and M. S. Valipour, "A new design of induced-charge electrokinetic micromixer with corrugated walls and conductive plate installation," *Int. Commun. Heat Mass Transf.*, vol. 114, no. March, p. 104564, 2020, doi: 10.1016/j.icheatmasstransfer.2020.104564.

28. S. Azimi, M. Nazari, and Y. Daghighi, "Developing a fast and tunable micromixer using induced vortices around a conductive flexible link," *Phys. Fluids*, vol. 29, no. 3, 2017, doi: 10.1063/1.4975982.

29. C. K. Harnett, J. Templeton, K. A. Dunphy-Guzman, Y. M. Senousy, and M. P. Kanouff, "Model based design of a microfluidic mixer driven by induced charge electroosmosis," *Lab Chip*, vol. 8, no. 4, pp. 565–572, 2008, doi: 10.1039/b717416k.

30. M. K. D. Manshadi, H. Nikookar, M. Saadat, and R. Kamali, "Numerical analysis of non-uniform electric field effects on induced charge electrokinetics flow with application in micromixers," *J. Micromechanics Microengineering*, vol. 29, no. 3, p. 035016, Feb. 2019, doi: 10.1088/1361-6439/AAFDC9.

Chapter 18

Numerical study of unsteady flow of hybrid nanofluid induced by a slendering surface with suction and injection effects

Moh Yaseen and Manoj Kumar

G.B. Pant University of Agriculture and Technology, Pantnagar, India

Sawan Kumar Rawat

Central University of Rajasthan, Ajmer, India

CONTENTS

18.1 INTRODUCTION

Occurrence of heat transfer is one of the main concerns in thermal and engineering systems. It is a significant phenomenon that exists between the two mediums due to a temperature difference within them. The efficient way to advance the thermal efficiency of a system is achieved by suspending solid nano-sized particles in the fluid. A uniform distribution of nano-sized particles in a conventional working fluid is known as nanofluid. A good thermal performance of a new generation fluid, known as a "hybrid nanofluid", is of assistance to accomplish the manufacturing or scientific needs. Hybrid nanofluid is a base fluid (i.e., water, ethylene, etc.) with two or more different types of nanoparticles (i.e., metals, MoS_2, SiO_2, metal oxides, etc.).

DOI: 10.1201/9781003257691-18

Flow analysis over irregular surface thickness is a novel research field because of the vital applications in industries dealing with manufacturing, acoustic segments, nuclear reactor engineering, etc. The researchers [1–3] have studied the flow problems over slendering surface in the past. The researchers [4–8] have studied the heat transfer phenomenon in the presence of nanoparticles.

18.1.1 Objective of the study

Based on the literature aforementioned, this chapter's aim is to analyze the suction and injection effects on unsteady flow of "SiO_2–MoS_2/water hybrid nanofluid" persuaded by a slendering surface situated in a porous medium. The previous reports ([1, 2]) on slendering surface can be extended for some effects of suction or injection, natural convention, porous medium, heat source/sink, and viscous dissipation. The flow and heat transfer problem are solved via MATLAB in-built function "bvp4c solver". Graphs are displayed to show the variation of different solution profiles due to parameters involved.

18.2 MATHEMATICAL MODEL

Consider the unsteady 2D hybrid nanofluid magnetohydrodynamics (MHD) convective flow over a slendering surface implanted in a porous medium (Figure 18.1). The stretching velocity of surface is denoted as

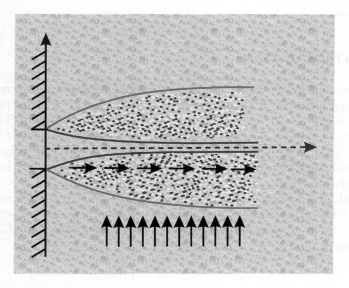

Figure 18.1 Physical layout of the flow.

$u_w = u_0 C^n (1 - mt)^{-1}$ and $v_w = \sqrt{v_0 C^{n-1}(1-mt)^{-1}}$ is the mass flux velocity, where $C = (x + a)$. Also, the starting point of flow is represented by $y = A\sqrt{(1 - mt)(x + a)^{1-n}}$, where A has a low order value, as the surface has non-uniform thickness. As shown in Figure 18.1, a transverse uniform magnetic field $B(x,t) = B_0 (C^{n-1}(1-mt)^{-1})^{\frac{1}{2}}$ is applied in the y-direction. Moreover, influence of Joule heating, thermal radiation, space-temperature reliant heat sink/source, viscous dissipation, and suction/injection effect is considered. In the present study, $n = 1$ represents the flat surface and $n = 0.5$ shows the flow on the slendering surface. Spherical-shaped nanoparticles are considered in this study.

The governing equations for the hybrid nanofluid flow are given by (see Ref. [2]):

$$u_y + v_y = 0 \tag{18.1}$$

$$
\begin{aligned}
u_t + u(u_x) + v(u_y) = {} & \frac{1}{\rho_{hbnf}} (\mu_{hbnf} u_{yy}) \\
& - \frac{1}{\rho_{hbnf}} (\sigma_{hbnf} B^2 u) \\
& + \frac{g(\rho\beta)_{hbnf}}{\rho_{hbnf}} (T - T_\infty) \\
& - \frac{\mu_{hbnf}}{\rho_{hbnf} k_0} u
\end{aligned}
\tag{18.2}
$$

$$
\begin{aligned}
\left(T_t + u(T_x) + v(T_y) \right) = {} & \alpha_{hbnf} (T_{yy}) \\
& - \frac{(q_r)_y}{(\rho C_p)_{hbnf}} \\
& + \frac{q'''}{(\rho C_p)_{hbnf}} \\
& + \frac{\mu_{hbnf} (u_y)^2}{(\rho C_p)_{hbnf}} \\
& + \frac{\sigma_{hbnf}}{(\rho C_p)_{hbnf}} B^2 u^2
\end{aligned}
\tag{18.3}
$$

The appropriate boundary conditions according to the discussion are as follows:

$$u(x,D) = u_w(x) + k_1 \times (u_y), \; v(x,D) = v_w(x) \atop T(x,D) = T_w(x) + k_2 \times (T_y), \; T \to 0, \; u \to 0 \text{ as } y \to \infty \bigg\} \tag{18.4}$$

where,

$$D = A\sqrt{\frac{(a+x)^{1-n}}{1-mt}},$$

$$k_1^* = \left(\frac{2-g_1}{g_1}\right)\zeta_1\sqrt{\frac{(b+x)^{1-n}}{1-ct}},$$

$$k_2^* = \left(\frac{2-a_1}{a_1}\right)\sqrt{\frac{(x+b)^{1-n}}{1-ct}},$$

$$\zeta_2 = \left(\frac{2\gamma_1}{\gamma_1+1}\right)\frac{\zeta_1}{\text{Pr}}$$

$$\tag{18.5}$$

In this chapter, the magnetic field is $B(x,t) = B_0(1-mt)^{-1/2}(x+a)^{\frac{n-1}{2}}$ and the temperature at the surface is noted as $T_w(x,t) = T_0(x+a)^{\frac{1-n}{2}}(1-mt)^{-1/2} + T_\infty$.

where velocity component along the "x—direction is u and velocity component along the y—direction is v" and v_w be the constant mass flow velocity. Furthermore, α_{hbnf}—thermal diffusivity, $C_{p_{hbnf}}$—heat capacity, ρ_{hbnf}—density, μ_{hbnf}—dynamic viscosity, k_{hbnf}—thermal conductivity, σ_{hbnf}—electrical conductivity, g—gravitational acceleration, β_{hbnf}—thermal expansion coefficient, T—temperature, T_∞—ambient temperature, B—variable magnetic field. Moreover, subscript $hbnf$—hybrid nanofluid, bf—base fluid, f—fluid (water).

Thus, expression for the term q''' is written as (see Ref. [8]):

$$q''' = \frac{k_f u_w(x,t)(T_w - T_\infty)}{(x+b)v_f}\left\{A^*\frac{dg}{\partial\xi} + B^*\theta\right\} \tag{18.6}$$

We have utilized the "linear Rosseland approximation for the radiation heat flux (q_r)" and the expression for the same is (see Ref. [9]):

$$q_r = -\frac{\partial T^4}{\partial y}\frac{4\sigma^*}{3k^*} \approx -\frac{\partial T}{\partial y}\frac{16T_\infty^3\sigma^*}{3k^*} \tag{18.7}$$

where "σ^* and k^* stand for Stefan–Boltzmann constant and mean absorption coefficient, respectively.

18.3 SIMILARITY TRANSFORMATION

We introduce the following similarity variable in governing equations, (see Ref. [2]).

$$
\left.\begin{aligned}
u &= \frac{u_0(x+a)^n g'}{1-mt}, \quad v = -\left(\xi g'\left(\frac{n-1}{n+1}\right) + g\right)\sqrt{0.5\nu U_0 \frac{(x+a)^{n-1}}{1-mt}(n+1)} \\
\theta(\eta) &= \frac{T - T_\infty}{T_w - T_\infty}, \quad \xi = y\sqrt{0.5\frac{U_0}{\nu}\frac{(x+a)^{n-1}}{1-mt}(1+n)},
\end{aligned}\right\}
\tag{18.8}
$$

Equations (18.2) and (18.3) are transformed utilizing the above Equation (18.8) as:

$$
\begin{aligned}
g''' = \frac{2}{n+1}\bigg(& \frac{\chi_B}{\chi_A}(0.5A(2g'+\eta g'') \\
& + ng'^2 - \frac{1}{2}(n+1)gg'') + \frac{\chi_C}{\chi_A}Mg' \\
& - \frac{\chi_D}{\chi_A}\gamma\theta + pg' \bigg)
\end{aligned}
\tag{18.9}
$$

$$
\begin{aligned}
\left(\frac{k_{hbnf}}{k_f} + R\right)\theta'' = \chi_E \Pr\bigg(& \frac{A}{n+1}(\theta+\eta\theta') + \left(\frac{1-n}{n+1}\right)g'\theta - \theta'g \bigg) \\
& - \left(\frac{2}{n+1}\right)[A^*g' + B^*\theta] \\
& - \chi_A \Pr Ec g''^2 - \chi_C Ec M \Pr\left(\frac{2}{n+1}\right)g'^2
\end{aligned}
\tag{18.10}
$$

Boundary conditions (1.4) becomes:

$$
\begin{aligned}
g(\varepsilon) &= S + \xi\left(\frac{1-n}{1+n}\right)[1 + k_1 g''(\varepsilon)], \quad g'(\varepsilon) = 1 + k_1 g''(\varepsilon), \\
\theta(\varepsilon) &= 1 + k_2\theta'(\varepsilon) \text{ at } \varepsilon = A\sqrt{\frac{(1+n)U_0}{2\nu}} \\
g'(\varepsilon) &= \theta(\varepsilon) = 0 \qquad\qquad \text{as } \varepsilon \to \infty
\end{aligned}
\tag{18.11}
$$

Where $k_1\left(=\zeta_1\left(\frac{2-g_1}{g_1}\right)\sqrt{(n+1)\frac{U_0}{2\nu}}\right)$—velocity slip parameter and $k_2\left(=\zeta_2\left(\frac{2-d_1}{d_1}\right)\times\sqrt{(n+1)\frac{U_0}{2\nu}}\right)$—temperature jump parameter. The non-dimensional parameter namely, $S\left(=-\sqrt{\dfrac{2v_0}{(n+1)vU_0}}\right)$—injection or suction

parameter, $A\left(=\dfrac{m}{U_0(x+a)^{n-1}}\right)$—unsteadiness parameter, $P\left(=\dfrac{(1-mt)v_f}{k_0 U_0(x+a)^{n-1}}\right)$

—porosity parameter, $M\left(=\dfrac{\sigma_f B_0^{\ 2}}{\rho_f U_0}\right)$—magnetic field, $\gamma\left(=\dfrac{Gr_x}{Re_x^{\ 2}}\right)$—

mixed convection parameter, $Ec\left(=\dfrac{U_w^{\ 2}}{(T_w-T_\infty)(C_p)_f}\right)$—Eckert number,

$\Pr\left(=\dfrac{u_f(C_p)_f}{k_f}\right)$—Prandtl number, $R\left(=\dfrac{16\sigma^*T_\infty^3}{3k_f k^*}\right)$—Radiation parameter. Also,

$\chi_A=\dfrac{\mu_f}{\mu_{hbnf}}, \chi_B=\dfrac{\rho_{hbnf}}{\rho_f}, \chi_C=\dfrac{\sigma_{hbnf}}{\sigma_f}, \chi_D=\dfrac{(\rho\beta)_{hbnf}}{(\rho\beta)_f} \chi_E=\dfrac{(\rho C_p)_{hbnf}}{(\rho C_p)_f}$. The above

thermophysical correlations of the hybrid nanofluid can be referred from Refs. [9] and [10]. Furthermore, Devi and Devi [10] have shown that these correlations show close agreement with experimental data.

For solving Equations (18.9) to (18.11), we have used the following functions to change the domain form of the problem from $[\varepsilon,\infty)$ to $[0,\infty)$ as $f(\eta)=f(\xi-\varepsilon)=g(\xi)$, then (18.9) to (18.11) become:

$$
\begin{aligned}
f''' = \frac{2}{n+1}\Bigg(&\frac{\chi_B}{\chi_A}(0.5A(2f'+\eta f'') \\
&+nf'^2-\frac{1}{2}(n+1)ff'' \\
&+\frac{\chi_C}{\chi_A}Mf'-\frac{\chi_D}{\chi_A}\gamma\theta+pf'\Bigg)
\end{aligned}
\tag{18.12}
$$

$$
\theta''=\frac{\chi_E\Pr\left(\dfrac{A}{n+1}(\theta+\eta\theta')+\left(\dfrac{1-n}{n+1}\right)f'\theta-\theta'f\right)-\left(\dfrac{2}{n+1}\right)[A^*f'+B^*\theta]}{\left(\dfrac{k_{hbnf}}{k_f}+R\right)}\\[6pt]
\frac{-\chi_A\Pr Ecf''^2-\chi_C EcM\Pr\left(\dfrac{2}{n+1}\right)f'^2}{\left(\dfrac{k_{hbnf}}{k_f}+R\right)}
\tag{18.13}
$$

$$
\begin{aligned}
&f(\eta)=S+\eta\left(\frac{1-n}{1+n}\right)[1+k_1 f''(\eta)], \\
&f'(\eta)=1+k_1 f''(\eta), \theta(\eta)=1+k_2\theta'(\eta) \quad \text{at } \eta=0 \\
&f'(\eta)=\theta(\eta)=0 \qquad \text{as } \eta\to\infty
\end{aligned}
\tag{18.14}
$$

The "Nusselt number" and "local skin friction" are engineering quantities and defined as:

$$C_{fx} = \frac{2\tau_w}{\rho_f u_w^2} \text{ and } Nu_x = \frac{x(q_w + q_r)}{k_f(T_w - T_\infty)} \tag{18.15}$$

where the "heat flux" $(q_w + q_r)$ and the "surface shear stress" τ_w are as follows:

$$q_w + q_r = \left(k_{hnf} \frac{\partial T}{\partial y} + \frac{4\sigma^*}{3k^*} \frac{\partial T^4}{\partial y} \right)\bigg|_{y=0} \text{ and } \tau_w = \mu_{hnf}\left(\frac{\partial u}{\partial y} \right)\bigg|_{y=0} \tag{18.16}$$

Using Equations (18.9) and (18.15), the Equation (18.16) is rewritten as:

$$\begin{aligned} \sqrt{Re_x}C_f &= \chi_A\sqrt{2(n+1)}f''(0) \\ \frac{Nu_x}{\sqrt{Re_x}} &= -\left\{\left(\frac{k_{hbnf}}{k_f} + R\right)\right\}\sqrt{\frac{n+1}{2}}\theta'(0) \end{aligned} \right\} \tag{18.17}$$

where $Re_x = \dfrac{u_w(x+a)}{v_f}$ represents the "local Reynolds number".

18.4 RESULTS AND DISCUSSION

This problem considers the flow of "SiO_2–MoS_2/water hybrid nanofluid" past a slendering surface. Detailed numerical computations are done to observe the effect of various inclusive parameters on "SiO_2–MoS_2/water hybrid nanofluid" flow and the results are presented graphically. Furthermore, the suction is represented by values of $S > 0$, while injection is represented by values of $S < 0$. The respective values of non-dimensional parameters values are set as default: $\eta = 6$, $\gamma = 5$, $A = A^* = B^* = Ec = P = k_1 = k_2 = 0.1$, $n = 0.5$, $M = R = 1$, $\varphi_{SiO_2} = \varphi_{MoS_2} = 0.1$, $\lambda = 0.5$, and $Pr = 6.2$. In figures, the solid line signifies the solution for suction ($S = 1$), and the dashed line signifies the solution of injection ($S = -1$). To validate the results of the study, a

Table 18.1 Thermophysical properties values of base fluid and nanoparticles (see Ref. [9]).

	$\rho(Kg/m^3)$	$C_p(J/KgK)$	$k(W/mK)$	$\beta(1/K)$
Water	997.1	4179	0.613	21
MoS_2	5060	397.746	34.5	2.84×10^{-5}
SiO_2	2650	730	1.5	42.7

Table 18.2 Comparison of $g''(0)$ for various values of the magnetic parameter (M) when $Pr = R = 1, \eta = 8$, $n = 1, \varphi_{SiO_2} = \varphi_{MoS_2} = A = \gamma = P = A^* = B^* = Ec = 0.$

M	Mabood and Das [11]	Present results
0.25	−1.118035	−1.118042
1	−1.4142135	−1.414214
5	−2.4494897	−2.44949
100	−10.049875	−10.049876

comparative analysis with results from published research of Mabood and Das [11] is done. A close precision is obtained in the comparison, as shown in Table 18.2.

18.4.1 Velocity profile

The change in the velocity profiles with changing the values of unsteadiness parameter (A) is sketched in Figure 18.2. The velocity increases while increasing the unsteadiness parameter (A) value in both injection and suction case. The increment in unsteadiness parameter (A) increases the movement of particles in the flow region; hence, an increase in the velocity is witnessed.

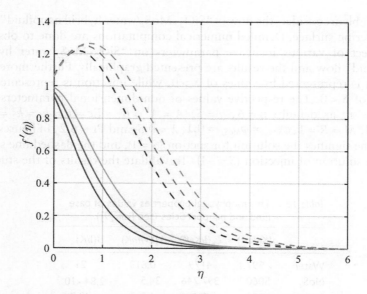

Figure 18.2 Influence of varying unsteadiness parameter (A) on $f'(\eta)$.

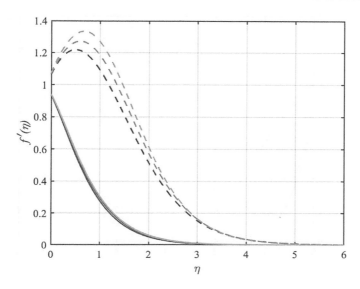

Figure 18.3 Influence of varying power-law index parameter (n) on $f'(\eta)$.

The variation in velocity due to power-law index (n) for both suction and injection cases is plotted in Figure 18.3. From the graph, it is apparent that the rise in velocity is seen with a rise in the power-law index (n) in both cases.

18.4.2 Temperature profile

Figure 18.4 elucidates the control of Eckert number (Ec) on hybrid nanofluid temperature during flow. The increment in temperature profile is seen, when Eckert number rises. This is because of existence of "viscous dissipation" term in the "energy equation", which physically represents self-heating of the fluid.

Figures 18.5 and 18.6 depict the temperature variation due to effect of "heat generation/absorption parameters A^* and B^{*}", respectively. The positive values of "heat source parameters A^* and B^* corresponds to the situation of heat generation within the system during the flow". In both suction and injection case, temperature rises with rising numerical values of A^* and B^*. Hence, temperature increases with generation of heat within flow region.

Figures 18.7–18.10 elucidate the impact of "magnetic parameter (M) and radiation parameter (R)" on "skin friction coefficient" and "heat transfer coefficient" with the suction and injection effect.

Figures 18.7 and 18.8 show that, the "skin friction coefficient" rises with an augmentation in the radiation parameter (R) and magnetic parameter (M). These results imply that the surface drag force increases with increment

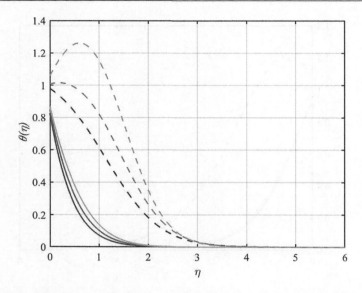

Figure 18.4 Influence of varying power-law index parameter (*n*) on $\theta(\eta)$.

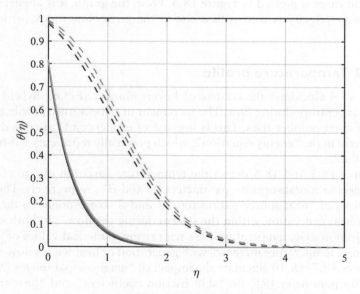

Figure 18.5 Influence of varying heat generation/absorption parameter (A^*) on $\theta(\eta)$.

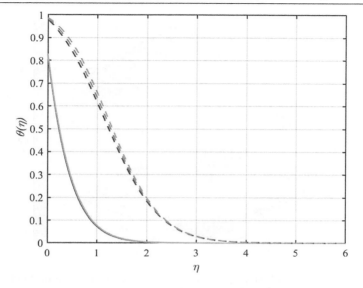

Figure 18.6 Influence of varying heat generation/absorption parameter (B^*) on $\theta(\eta)$.

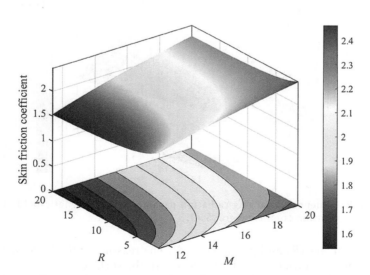

Figure 18.7 Influence of varying radiation parameter and magnetic field on skin friction coefficient for $S = 1$.

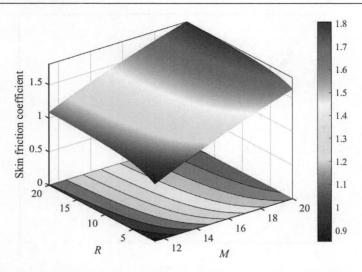

Figure 18.8 Influence of varying radiation parameter and magnetic field on skin friction coefficient for $S = -1$.

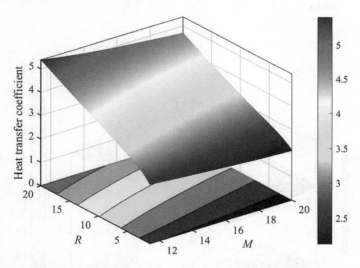

Figure 18.9 Influence of varying radiation parameter and magnetic field on heat transfer coefficient for $S = 1$.

in the parameter (R) and parameter (M). The reason for the increased drag force is increased friction of fluid layers with the surface.

Figures 18.9 and 18.10 show that, the "heat transfer coefficient" rises with an augmentation in the radiation parameter (R) but decreases with increment in magnetic parameter (M). The heat transport escalates with an increment in the parameter (R) due to increased radiation and the heat

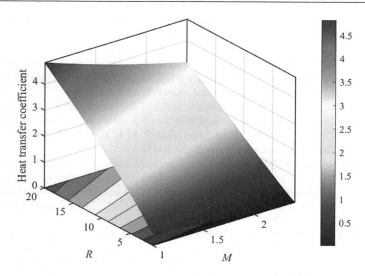

Figure 18.10 Influence of varying radiation parameter and magnetic field on heat transfer coefficient for $S = -1$.

transport decreases with an increment in the parameter (M) due to generation of the Lorentz force in the reverse direction of the flow.

18.5 CONCLUSIONS

The main observations are summarized as follows:

- The problem examines the injection and suction effects on the "hybrid nanofluid flow" over slendering surface.
- The thermal profile of the SiO_2–MoS_2/water hybrid nanofluid is enhanced by the increase in the Eckert number and the heat absorption parameters.
- Results point out that the flow with injection effect is more effective as compared to suction effect. In the case of the injection effect, the velocity and temperature are observed to be greater.
- Hybrid nanofluid has a dominant effect on temperature distribution.

REFERENCES

1. J.V. Ramana Reddy, V. Sugunamma, and N. Sandeep, "Thermophoresis and Brownian motion effects on unsteady MHD nanofluid flow over a slendering stretching surface with slip effects." *Alexandria Eng. J.* 57, 2465–2473, 2018. https://doi.org/10.1016/j.aej.2017.02.014

2. F. Mabood, and G.P.A.N. Sandeep, "Simultaneous results for unsteady flow of MHD hybrid nanoliquid above a flat/slendering surface." *J. Therm. Anal. Calorim.* 2020. https://doi.org/10.1007/s10973-020-09943-x

3. T. Hayat, M. Farooq, A. Alsaedi, and F. Al-Solamy, "Impact of Cattaneo-Christov heat flux in the flow over a stretching sheet with variable thickness." *AIP Adv.* 5, 0–12, 2015. https://doi.org/10.1063/1.4929523

4. R.U. Haq, A. Raza, E.A. Algehyne, and I. Tlili, "Dual nature study of convective heat transfer of nanofluid flow over a shrinking surface in a porous medium." *Int. Commun. Heat Mass Transf.* 114, 104583, 2020. https://doi.org/10.1016/j.icheatmasstransfer.2020.104583

5. G. Rasool, and A. Wakif, "Numerical spectral examination of EMHD mixed convective flow of second-grade nanofluid towards a vertical Riga plate using an advanced version of the revised Buongiorno's nanofluid model." *J. Therm. Anal. Calorim.* 143, 2379–2393, 2021. https://doi.org/10.1007/s10973-020-09865-8

6. H. Upreti, A.K. Pandey, and M. Kumar, "MHD flow of Ag-water nanofluid over a flat porous plate with viscous-Ohmic dissipation, suction/injection and heat generation/absorption." *Alexandria Eng. J.* 57, 1839–1847, 2018. https://doi.org/10.1016/j.aej.2017.03.018

7. F.A. Soomro, R.U. Haq, Q.M. Al-Mdallal, and Q. Zhang, "Heat generation/absorption and nonlinear radiation effects on stagnation point flow of nanofluid along a moving surface." *Results Phys.* 8, 404–414, 2018. https://doi.org/10.1016/j.rinp.2017.12.037

8. N. Sandeep, and M.G. Reddy, "Heat transfer of nonlinear radiative magnetohydrodynamic Cu-water nanofluid flow over two different geometries." *J. Mol. Liquids* 225, 87–94, 2017. https://doi.org/10.1016/j.molliq.2016.11.026

9. M. Yaseen, S.K. Rawat, and M. Kumar, "Hybrid nanofluid (MoS2–SiO2/water) flow with viscous dissipation and Ohmic heating on an irregular variably thick convex/concave-shaped sheet in a porous medium." *Heat Transfer*, 2021. https://doi.org/10.1002/htj.22330

10. S.S.U. Devi, and S.A. Devi, "Numerical investigation of three-dimensional hybrid Cu–Al2O3/water nanofluid flow over a stretching sheet with effecting Lorentz force subject to Newtonian heating." *Can. J. Phys.*, 94, 490–496, 2016.

11. F. Mabood, and K. Das, "Melting heat transfer on hydromagnetic flow of a nanofluid over a stretching sheet with radiation and second-order slip." *Eur. Phys. J. Plus*, 131, 1–12, 2016.

Chapter 19

Computational analysis of toe-out type vortex generators for improved thermal capacity of finned tube arrays

Amit Arora

Malaviya National Institute of Technology Jaipur, Jaipur, India

CONTENTS

19.1 INTRODUCTION

Thermal resistance in heat exchange systems is being reduced as much as feasible. As a passive heat transfer enhancement strategy, swirl flow generators can be incorporated into the base flow [1]. As a result, they form three-dimensional longitudinal vortices that promote bulk mixing in the base flow. The heat exchange module's unit cost is reduced as a result of the flow improvements, which allow for higher heat transfer coefficients [2]. A finned tube array with vortex generators solely behind the first row of tubes resulted in a 16.5–44% improvement in heat transfer coefficient [3]. Deployment of generators behind each alternate row of tubes further increased the heat transmission coefficient by 30–68.8%. Likewise, the importance of generator positioning and its geometric design was emphasized in another study that conducted computational investigation to analyze the two aspects [4]. Another computational study talks about the significance of the attack angle of the vortex generators [5], and it is reported that both Nusselt number and pressure drop increase with the attack angle. Because of manufacturing constraints, some flexibility in the spatial location of the vortex generators downstream is required to attain the desired thermal augmentation. Since the usefulness of a pair of vortex generators varies depending on where they

DOI: 10.1201/9781003257691-19

are situated, it is important to place them in a strategic location. A study of the impact of varying LVG position on wake control is needed in order to take advantage of vortex generators for thermal augmentation in finned tube arrays. Flow symmetry in the cross-stream direction necessitates this study to examine the influence of cross-stream translation on flow altera- tions and heat transfer enhancement. In spite of the ease of making rectan- gular winglets, delta-winglet vortex generators (DWLVG) are preferred due to their superior performance over rectangular ones [6].

19.2 NUMERICAL SIMULATIONS

A two-equation model Renormalization Group (RNG k-ε) for turbulence modeling is used for simulating the flow mechanics [2, 7], and the flow is assumed to be a steady incompressible turbulent flow. The governing equa- tions can be found in the open literature [8] with ease. The discretization of the governing equations is accomplished via the use of the finite volume method [9], and the pressure–velocity coupling is accomplished through the use of the Semi-Implicit Method for Pressure-Linked Equations (SIMPLE) method [10]. Conjugate heat transfer has been suggested as a way to predict heat exchange through the fins. The tube wall is assumed to be isother- mal [2, 11]. In accordance with a previously reported experimental inves- tigation [12], a fin-tube array with three tube rows in the flow direction is simulated, as shown in Figure 19.1. Furthermore, the solution domain is periodic in the normal direction and symmetric in the transverse direction. Flow lengths equaling 10S and 30S are also specified at the inlet and exit, respectively [12]. The vortex generators are set up with a 45° attack angle, and the boundary conditions are shown in Figure 19.1.

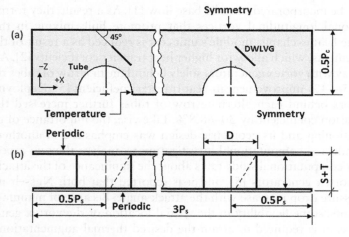

Figure 19.1 Solution domain: (a) plan and (b) elevation.

19.2.1 Model validation

The numerical predictions are compared with published experimental data for model validation [12], as shown in Figure 19.2. The friction factor is represented on the y-axis, while the Reynolds number is represented on the x-axis. The hydraulic diameter is chosen as the characteristic dimension, and the far-field inflow velocity is treated as the characteristic velocity [13, 14].

19.3 RESULTS AND DISCUSSION

As illustrated in Figure 19.3, this analysis identifies three distinct streamwise sites for erecting the generators (i.e., X/D = 0.3, 0.7, and 1.1). The generators are transversely displaced. Evidently, three distinct cross-stream sites correspond to each streamwise position of DWLVG. The Reynolds number (Re = 4245) is maintained fixed, the value equals 4245, with the winglets positioned in toe-out orientation.

19.3.1 Change in flow structure

Emphasis has been given to the velocity vectors in order to get insight into the generated secondary flow structures. The vectors corresponding to a particular location of the generators are presented in Figures 19.4(a) and 19.4(b). The leading plane intercepts the generators, whereas the following plane is placed in the generators' downstream.

There is a vortical formation of dismal flow velocities, which evolves in the flow direction. Such a structure may be attributed to sudden flow separation on account of steep attack angle. In addition, there is a prominent flow diversion toward the leeward side of the tubes due to toe-out erection

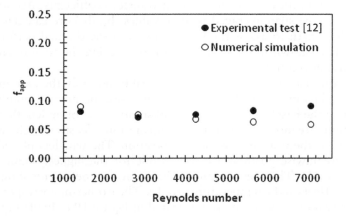

Figure 19.2 Friction factor variation.

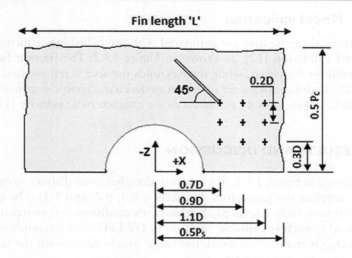

Figure 19.3 DWLVG positions in the tube downstream.

of the generators. The change in the flow structures caused due to the shifting of generators is presented in Figure 19.4(c). Since the generators, placed at Z/D=0.3, stand behind the tubes, the vortical formation is gloomy. However, as the generators shift outward, the said formation is apparent. In fact, the same goes for the flow diversion caused by the generators.

19.3.2 Thermal characteristics

Once the flow changes have been visually assessed, it is customary to inquire about the benefits accruing in terms of heat transfer enhancement. Despite the fact that the Reynolds number remained unchanged, flow changes result in significant changes in the heat transfer coefficients. According to Figure 19.5, the span averaged relative values (h_{VG}/h_o) are calculated over the full fin length to assess the local changes in the thermal performance. The x-axis is shown in dimensionless form by using the tube pitch in the streamwise direction (P_s).

For a contrasting perspective of the distribution, only the extreme positions are taken into consideration. In both cases, the Nusselt numbers are higher than the baseline values over a substantial part of fin length, and the distributions are mostly similar. The augmentation is least when the winglets are placed at the nearest cross-stream position. The winglets placed at the leading streamwise position are responsible for the highest augmentation ratio of 206.6%. The augmentation in heat transfer coefficients should have a marked effect on the temperature profiles. The isotherms corresponding to leading streamwise position are illustrated in Figure 19.6. Evidently, the fin is definitely benefited irrespective of the generators' placement. However,

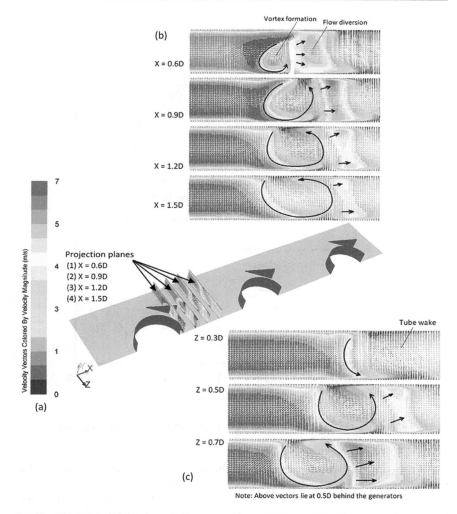

Figure 19.4 Velocity vectors: (a) modified fin, (b) secondary flow generation by DWLVGs placed at X = 0.7D, Z = ±0.5D, and (c) cross-stream translation of DWLVG at X = 0.9D.

the temperatures are declining with the shifting of generators in the transverse direction.

Figure 19.7 shows the effect of winglet translation on the average heat transfer coefficient. With the generators' outward cross-stream translation, the relative values clearly rise. Although the trend remains the same regardless of winglet position in the streamwise direction, the degree of augmentation decreases as winglets undergo streamwise translation. The relative heat transfer coefficient for the three streamwise positions attains the highest value of 128.5, 126.5, and 124.5%, respectively.

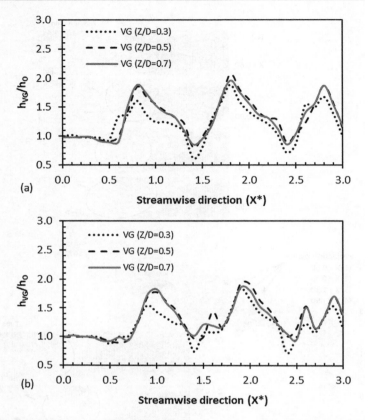

Figure 19.5 Heat transfer coefficient for fixed DWLVG position (a) X/D = 0.7 and (b) X/D = 1.1.

All tubes and a considerable part of the fins, especially which is wetted by the wakes, make up the wake-affected surfaces in a fin-tube array. By looking at the velocity contours, as shown in Figure 19.8(a), the spread of the latter surface can be approximated. Figures 19.9 and 19.10 show the effect of winglet translation on the relative Nusselt numbers corresponding to the wake-affected fin and tubes, respectively. Evidently, heat transfer augmentation over the said fin surface attains maximum value at the intermediate cross-stream position. Such a trend suggests that the generators should be neither too close to the tubes nor too far. For the three streamwise positions, the highest augmentation ratio over the wake-affected fin is 214.9, 200.2, and 184.3%, respectively. Unlike fins, the trend of augmentation is monotonic in case of tubes with nearest cross-stream location being counterproductive. The augmentation over leading tube is virtually insensitive to the generators' position.

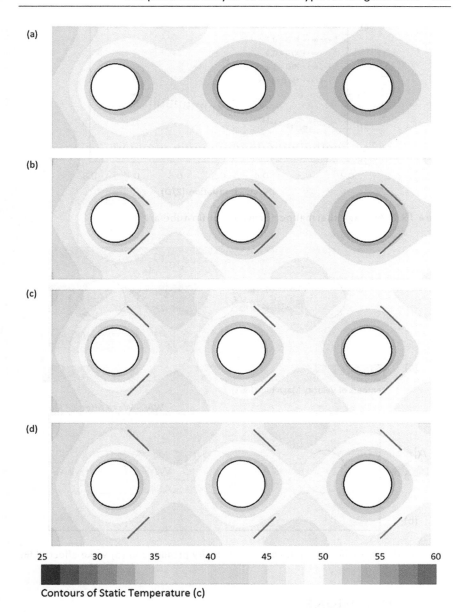

Contours of Static Temperature (c)

Figure 19.6 Isotherms (a) No DWLVG, (b) Z/D = 0.3, (c) Z/D = 0.5, and (d) Z/D = 0.7.

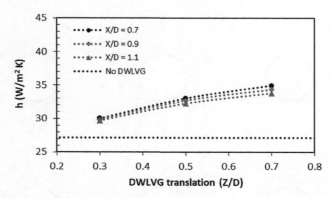

Figure 19.7 Average thermal performance of fin-tube array.

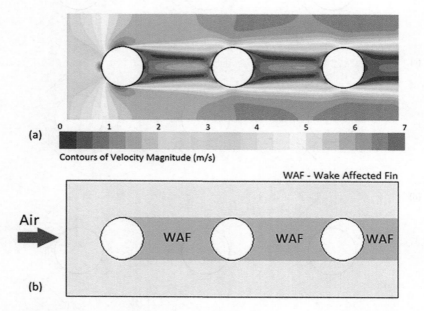

Figure 19.8 Wake-affected fin (a) baseline velocity profiles and (b) wake-affected fin.

19.4 CONCLUSIONS

Heat transfer augmentation in finned tube arrays is achieved using delta type vortex generators in this work. The generators are erected with toe-out orientation, and they are deployed behind each tube row. The impact of generator cross-stream translation on flow structures and heat transfer

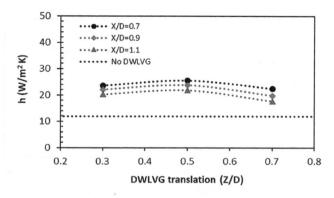

Figure 19.9 Heat transfer coefficient of the wake-affected fin.

enhancement is presented in this computational study. Regardless of the streamwise position, the average convection coefficient rises with their translation in the outer direction. However, in order to provide optimum heat transfer enhancement across the wake-affected fin, the winglets should be neither too close nor too distant from the tubes. The highest thermal augmentation ratio for the wake-affected fin is 153.2%, and its overall average value is 214.9%.

NOMENCLATURE

D	Tube diameter	m
LVG	Longitudinal vortex generators	–
f_{app}	Friction factor	–
h	Heat transfer coefficient	W/m² K
L_c	Characteristic length	m
P_c	Cross-stream pitch	m
P_s	Streamwise pitch	m
Re	Reynolds number	–
S	Fin space	m
T	Fin thickness	m
X	Streamwise coordinate	m
Z	Cross-stream coordinate	m
ρ	Fluid density	kg/m³

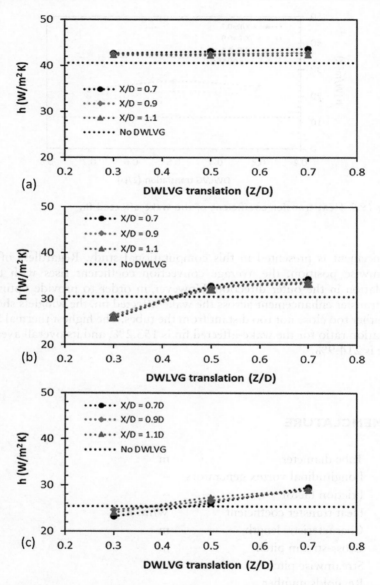

Figure 19.10 Heat transfer augmentation on the tubes (a) Tube 1, (b) Tube 2, and (c) Tube 3.

REFERENCES

1. R.L. Webb, and N.H. Kim, *Principles of Enhanced Heat Transfer*. 2nd edition, Taylor & Francis, New York, 2005.
2. M. Gorji, H. Mirgolbabaei, A. Barari, G. Domairry, and N. Nadim, "Numerical analysis on longitudinal location optimization of vortex generator in compact

heat exchangers." *International Journal for Numerical Methods in Fluids*, 66, pp. 705–713, 2011.

3. A. Joardar, and A.M. Jacobi, "Heat transfer enhancement by winglet-type vortex generator arrays in compact plain-fin-and-tube heat exchangers." *International Journal of Refrigeration*, 31, pp. 87–97, 2008.

4. H. Huisseune, C. T'Joen, P.D. Jaeger, B. Ameel, S.D. Schampheleire, and M.D. Paepe, "Influence of the louver and delta winglet geometry on the thermal hydraulic performance of a compound heat exchanger." *International Journal of Heat and Mass Transfer*, 57, pp. 58–72, 2013.

5. A. Pal, D. Bandyopadhyay, G. Biswas, and V. Eswaran, "Enhancement of heat transfer using delta-winglet type vortex generators with a common-flow-up arrangement." *Numerical Heat Transfer: Part A*, 61, pp. 912–928, 2012.

6. J.M. Wu, and W.Q. Tao, "Numerical study on laminar convection heat transfer in a channel with longitudinal vortex generator. Part B: Parametric study of major influence factors." *International Journal of Heat and Mass Transfer*, 51, pp. 3683–3692, 2008.

7. G. Lu, and G. Zhou, "Numerical simulation on performances of plane and curved winglet type vortex generator pairs with punched holes." *International Journal of Heat and Mass Transfer*, 102, pp. 679–690, 2016.

8. A. Arora, P.M.V. Subbarao, and R.S. Agarwal, "Numerical optimization of location of 'common flow up' delta winglets for inline aligned finned tube heat exchanger." *Applied Thermal Engineering*, 82, pp. 329–340, 2015.

9. D.J. Dezan, L.O. Salviano, and J.I. Yanagihara, "Interaction effects between parameters in a flat-tube louvered fin compact heat exchanger with delta-winglets vortex generators." *Applied Thermal Engineering*, 91, pp. 1092–1105, 2015.

10. J. Gong, C. Min, C. Qi, E. Wang, and L. Tian, "Numerical simulation of flow and heat transfer characteristics in wavy fin-and-tube heat exchanger with combined longitudinal vortex generators." *International Communications in Heat and Mass Transfer*, 43, pp. 53–56, 2013.

11. A. Joardar, and A.M. Jacobi, "A numerical study of flow and heat transfer enhancement using an array of delta-winglet vortex generators in a fin-and-tube heat exchanger." *Journal of Heat Transfer*, 129, pp. 1156–1167, 2007.

12. A. Arora, P.M.V. Subbarao, and R.S. Agarwal, "Development of parametric space for the vortex generator location for improving thermal compactness of an existing inline fin and tube heat exchanger." *Applied Thermal Engineering*, vol. 98, pp. 727–742, 2016.

13. K.M. Kwak, K. Torii, and K. Nishino, "Simultaneous heat transfer enhancement and pressure loss reduction for finned-tube bundles with the first or two transverse rows of built-in winglets." *Experimental Thermal and Fluid Science*, vol. 29, pp. 625–632, 2005.

14. A. Sinha, H. Chattopadhyay, A.K. Iyengar, and G. Biswas, "Enhancement of heat transfer in a fin-tube heat exchanger using rectangular winglet type vortex generators." *International Journal of Heat and Mass Transfer*, vol. 101, pp. 667–681, 2016.

Index

For Product Safety Concerns and Information please contact our
EU representative GPSR@taylorandfrancis.com Taylor & Francis
Verlag GmbH, Kaufingerstraße 24, 80331 München, Germany